QUESTIONS AND ANSWERS IN ELECTRICAL INSTALLATION WORK

QUESTIONS AND ANSWERS IN ELECTRICAL INSTALLATION WORK

R. A. MEE
M.I.T.E., M.A.S.E.E., ASSOC.M.C.T.
Openshaw Technical College, Manchester

MACDONALD : LONDON

Published in 1971 by Macdonald & Co. (Publishers) Ltd
St. Giles House, 49/50 Poland Street, London, W.1

SBN 356 03505 0

PRINTED IN UNITED KINGDOM BY
W. & G. BAIRD LTD.,
LONDON AND BELFAST

PREFACE

This book has been written with a practical approach to the techniques of electrical installation, some of which may not be familiar to the trained electrician or engineer.

Much of the theoretical fundamentals have been omitted in order to concentrate more on installation practice and techniques which do not normally appear in the conventional textbook.

My thanks are due to those of my colleagues and students who have kindly assisted me, and also to various firms who have allowed me to publish illustrations of their products. These include Crabtree Ltd., English Electric Ltd., Bill Switchgear Ltd., Gent & Co. Ltd., Johnson & Phillips, BICC, W. T. Henley's Telegraph Works & Co. Ltd., Walsall Conduits Ltd., Meritus (Barnet) Ltd., Evershed & Vegnoles Ltd., Foster Transformers Ltd., Hackbridge and Hewittic Electric Co. Ltd., Crompton Parkinson Ltd., Belling-See Ltd.

October 1971 R. A. Mee

CONTENTS

	Page
1. Consumer's terminals	1
2. Consumer's circuits	15
3. Cables	23
4. Wiring systems	38
5. Wiring systems in special situations	79
6. Earthing	98
7. Testing	115
8. Transformers	140
9. D.C. supplies	154
10. Power factor	167
11. Instruments	171
12. D.C. motors	186
13. A.C. motors	200
14. Illumination	230
15. Heating	242

1

Consumer's terminals

Domestic installations—miniature circuit-breakers—semi-enclosed and h.b.c. fuses—medium-sized industrial installations—large industrial installations—substations, fuse-switches, air and oil-filled circuit-breakers—discrimination.

1. *Describe the construction and operation of a miniature circuit-breaker and state what its advantages are when compared to the fuse.*

The miniature circuit-breaker, as its name suggests, is small and compact. The type illustrated in Fig. 1 has a magnetic-hydraulic time-delay, tripping mechanism consisting of a hermetically-sealed tube filled with a silicone fluid and

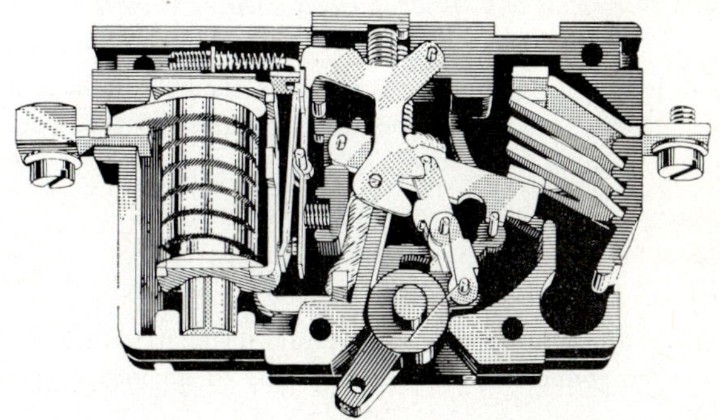

FIG. 1 *Miniature circuit-breaker.* (*Courtesy Crabtree Ltd.*)

containing a closely fitting iron core, all contained within an insulated case. Under normal load conditions, the magnetic pull established by the trip coil is not sufficient to overcome the force exerted by the time-delay spring, hence the core remains at the far end of the tube (Fig. 2).

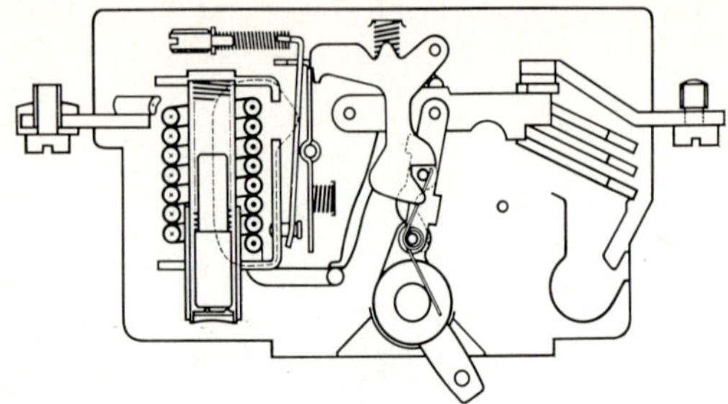

FIG. 2 *Operation of a miniature circuit-breaker.* (*Courtesy Crabtree Ltd.*)

Under moderate overload conditions, the magnetic pull causes the core to move along the tube, the speed of travel being related to the magnitude of the overload. If the overload is sustained, the pull on the armature is such that eventually the breaker is tripped (Fig. 3).

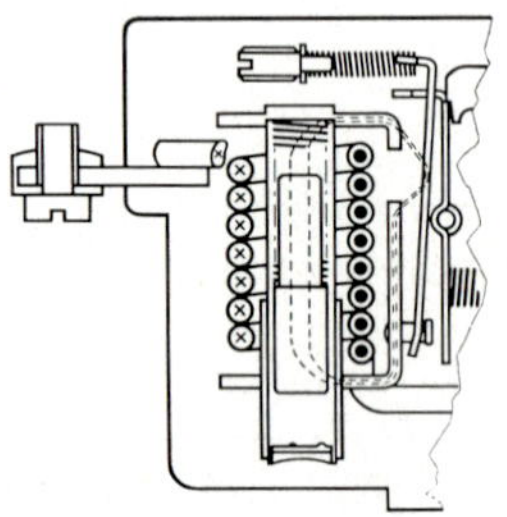

FIG. 3 *Operation of a miniature circuit-breaker.* (*Courtesy Crabtree Ltd.*)

For heavy overloads, or short-circuit conditions, the pull on the armature is so strong that the breaker trips instantaneously, the speed of the operation comparing favourably with the h.b.c. fuse.

The advantages of the miniature circuit-breaker are:

(1) It does give moderate excess current protection.

(2) It is far easier to replace than the fuse.

The latter advantage is important in domestic installations where the wrong size of fuse is often found installed, usually due to an inexperienced person replacing the fuse.

The disadvantages are:

(1) It is more expensive to install than the fuse, especially in domestic installations where the semi-enclosed fuse is normally installed.

(2) Occasionally the breaker trips inadvertently on normal loads, possibly through maladjustment.

2. *What are the fusing factor of a fuse and its category of duty?*

The fusing factor of a fuse is given as

$$\frac{\text{minimum fusing current}}{\text{fuse rating}}$$

It is now of great importance to know the fusing factor of a fuse for a particular duty as, under the latest edition of the IEE Regulations, it can affect the size of cable required for a given current loading. Also, in some installations it affects the value of the permissible impedance for the earth path.

The fusing factor for a semi-enclosed fuse is approximately two, which means that, while the fuse retains its correct rating, it will not break under approximately twice its rated current.

There are several types of h.b.c. fuse with different fusing factors, but from the electricians' point of view the important one is the one most commonly used in installation work, the class Q type. The fusing factor of this fuse marks the dividing line between coarse and close excess-current protection (see question **19**) and also the limits of the earth-loop impedance.

The Regulations also specify that the means of excess-current protection must be suitable for the maximum short-circuit current attainable. The category of duty is important, therefore, in that it is related to the prospective current of a circuit. Fuses are therefore divided into various categories of duty, dependent upon the prospective current attainable.

For final sub-circuits, where the prospective current is possibly less than 4000 A, a fuse with a category of duty AC2 or DC2, which includes the semi-enclosed fuse, would be adequate, whereas, at the intake position of a medium size installation, where the prospective current would possibly be in the order of 45 000 A, a h.b.c. fuse with a category of duty AC5 would be necessary.

3. *Describe the construction of a high-breaking capacity fuse and state what its advantages and disadvantages are when compared to the semi-enclosed fuse.*

The h.b.c. fuse, illustrated in Fig. 4, has a silver element contained in a sealed ceramic tube. The interior of the tube is filled with quartz sand so that there is no risk of ejection of hot metal when the fuse blows. At each end, heavy copper contacts are fitted to the tube, to which the element is connected. In many h.b.c. fuses, part of the element passes through small holes to the exterior of the tube, so giving visual indication when the fuse has blown.

The advantages of the h.b.c. fuse compared with the semi-enclosed fuse are:

(1) They have a much higher breaking capacity, being approximately 40 000 kVA for the h.b.c. fuse compared with 5000 kVA for the semi-enclosed fuse.

(2) Due to oxidation, the semi-enclosed fuse is subject to deterioration (that is, its cross-sectional area gradually diminishes), whereas the h.b.c. fuse is non-deteriorating.

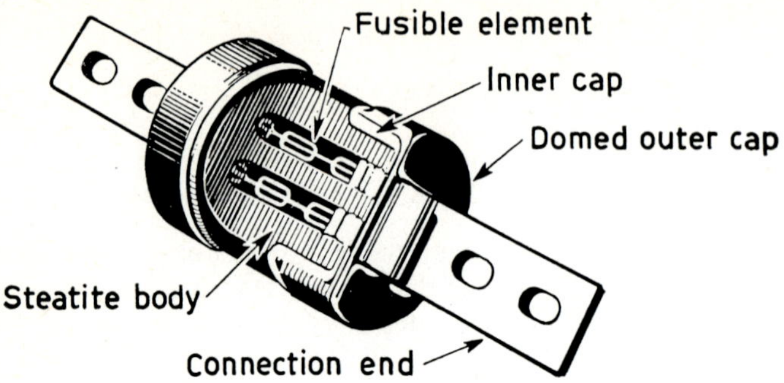

FIG. 4 *Construction of an h.b.c. fuse.* (*Courtesy English Electric Ltd.*)

(3) The h.b.c. fuse clears heavy faults with greater rapidity. This ability of a fuse to grade its breaking capacity time in inverse proportion to the current is termed the inverse characteristic of the fuse. If a h.b.c. fuse with a certain characteristic curve is selected, that curve will represent the performance of the fuse throughout its working life. It is this ability of the h.b.c. fuse to clear heavy faults rapidly, that makes it very suitable for the backing up of the slower acting circuit-breaker.

(4) Because of deterioration and also because the wrong size of element is so easily installed, discrimination is very poor where semi-enclosed fuses are installed, whereas where h.b.c. fuses are employed, it is possible to discriminate accurately between settings of the various protective devices throughout the entire system.

(5) It has a lower fusing factor (see question **2**).

(6) When the semi-enclosed fuse blows, there is always the danger of hot metal being scattered, especially where the asbestos tube has been removed. This constitutes a grave danger to operatives who are possibly replacing the fuse under short-circuit conditions, whereas, as the breaking of the fuse takes place in a sealed chamber in the h.b.c. fuse, this hazard is greatly reduced. The only disadvantage of the h.b.c. fuse is that it is more expensive to install and replace than the semi-enclosed fuse.

4. *What is meant by the* prospective current *of a circuit, and how does it affect the required breaking capacity of switchgear?*

The prospective current is the maximum current that could flow in the circuit if the protective device was replaced by a link with negligible impedance. The product of this current and the circuit voltage determine the breaking capacity of the switchgear.

In the layout shown in Fig. 5, it is clear that, because of the impedance of the mains and submains, the prospective current is at its lowest value at E and its highest at A, the switchgear controlling the installation. Thus, the breaking capacity of the switchgear must be graded according to its position in the distribution system, ranging from a breaking capacity of about 5 MVA at E to possibly 250 MVA at A, with intermediate values at B, C and D.

Another factor which must be taken into consideration when calculating the prospective currents is the reactance of alternator and transformers windings. The nearer the fault is to the transformer terminals, the less is the impedance in circuit; when the fault lies across the secondary winding terminals, the impedance in the circuit is virtually negligible and therefore, the reactance of the transformer is virtually the only limiting factor in the circuit.

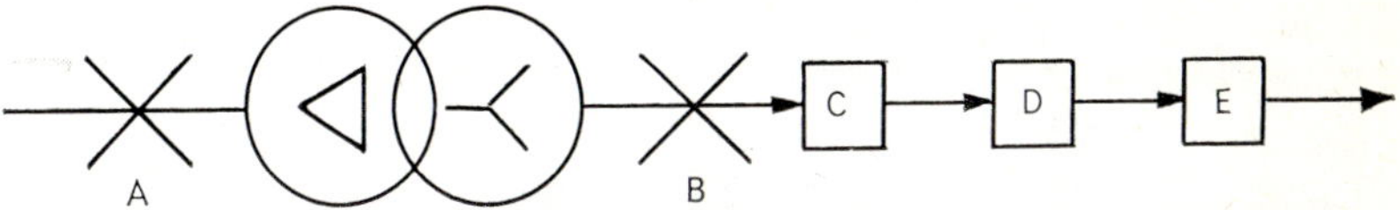

FIG. 5 *Illustrating the prospective current of circuits.*

It is important then to know the reactance values of the transformers in an installation, and this information is contained on the rating plate of the transformer. Its value is usually quoted as a percentage of the voltage drop at normal full load, thus

$$\%\ \text{reactance} = \frac{\text{fall in voltage due to full-load current}}{\text{open-circuit voltage}} \times 100$$

A further expression for the percentage reactance is

$$\%\ \text{reactance} = \frac{\text{full-load kVA}}{\text{short-circuit kVA}} \times 100$$

For a good transformer this is normally in the order of 5 per cent.

It is also clear from the second equation that the short-circuit kVA is equal to

$$\frac{\text{full-load kVA}}{\text{reactance}}$$

therefore, where the reactance of the transformer is 5 per cent, the short circuit kVA would be twenty times this value, that is a 1000 kVA transformer would have a short-circuit value of 20 000 kVA. It is clear then that the reactance of the windings plays an important part in determining the prospective current in the circuit and also the breaking capacity of the switchgear.

5. *Describe a suitable layout for the control equipment of a medium-size industrial installation.*

Fig. 6 illustrates a layout of the control gear for a medium-size industrial installation. It comprises an incoming, oil-filled circuit-breaker controlling a

A

busbar chamber on which are mounted the outgoing fuse-switches, each of which possibly controls a section or final distribution board.

The circuit-breaker controlling the triple-pole and neutral busbar chamber would be of the TP and N type. However, some of the outgoing fuse-switches would possibly be single-pole and neutral and be used for the single-phase lighting and power services, although in the illustration these services are shown connected across the three-phase and neutral busbar. Where SP and N supplies are used, these should be balanced across the three-phase and neutral busbars. The outgoing three-phase and neutral switch-fuses would be connected to the respective phase busbars and the neutral busbar.

FIG. 6 *Control gear for a medium size industrial installation.* (*Courtesy Bill Switchgear Ltd.*)

The Authority's equipment would also be incorporated, and as the consumer would probably be subject to a maximum demand tariff, an MDI meter would be installed, with possibly a kVAr meter where the consumer is subject to power factor penalty clauses.

6. *Describe the precautions that should be taken in the construction and layout of a sub-station to guard against risk of fire and leakage of oil associated with the use of transformers and electrical switchgear.*

The prime consideration has to be the elimination of all possible risk to personnel, and fire hazards from explosions or the leakage of hot oil from transformers and switchgear. Before construction work is started, it must be ensured that the proposed layout meets with the requirements of the local bye-laws, the Town and Country Planning Authorities and also satisfies the Power Regulations.

The floor should have a granolithic finish, and if the station does have a trench for catching escaping oil, the floor should slope slightly towards it. When finished, the floor should be treated with a special paint to reduce the accumulation of dust and to provide a non-slip surface.

The roof should be of concrete or a pre-cast concrete structure, and if steel joists are installed, they should be covered with concrete to prevent twisting.

The walls may be of a brick or cavity-brick construction. Where possible, it is preferable to contain the high- and medium-voltage switchgear, as well as the transformers, in separate rooms (in many cases, the transformers are situated outside the substation). No windows are necessary as most stations run unattended for long periods, and windows create an additional hazard in the event of an explosion. On the other hand, windows do serve to relieve pressure if explosions occur, and therefore, where they are omitted, unbonded panels may be built into unloaded walls for the same purpose. When finished, the walls should be painted or preferably tiled since this helps to stop dust accumulating and assists cleaning.

All doors should preferably be constructed of light steel; wood is a possible substitute, but where used, it should be painted with a good fire-resistant paint, as should all woodwork inside the station. Doors should always be arranged to open outwards. Two doors are necessary for each chamber, and at least one must be adequate for moving equipment in or out of the building. All emergency doors must be fitted with crash bolts, and all doors, with the exception of intercommunicating doors, should have robust locks (with keys) to prevent unauthorised entry.

In order to minimise condensation on walls and switchgear, ventilation must be adequate but not excessive. Better ventilation is required where the transformers are sited internally. Good practice is to allow 1 m^2 of free opening for every 30 m^3. This can be materially assisted by fitting inlets at low and high levels, the low level inlets sometimes being provided by louvres in external doors. As the ventilation system has to assist in cooling the transformer(s), these should not be placed too near to walls or other transformers. About 1·5 m of free air is necessary above the tank in order to assist in the circulation of air. Where the switchgear is placed in separate chambers, because of the low emission of heat, some form of artificial heating is required to maintain the temperature at about 10° C during cold weather.

High-voltage cables are normally installed in separate earthenware ducts terminating 75 mm to 100 mm above ground level. The layout of the ducts must be carefully routed and the radius of any bend must be adequate for the cables to be installed.

Medium-voltage cables are usually run in trenches although, in some cases, it may be more convenient to carry cables on cleats around the station walls. Where trenches are used, switchgear must not impede the easy drawing-in or withdrawal of cables, and must be in such a position as to facilitate cable jointing. Cables should also be supported on cleats or cable racks inside the trench instead of being laid untidily on the floor of the trench.

The Regulations specify that where oil-filled apparatus with an oil capacity exceeding 100 litres is installed, means must be provided for draining away escaping oil before it can reach other parts of the installation.

One method is to install a trench the full length of the switchgear of sufficient capacity to accommodate all the escaping oil. Small graded pebbles, but not sand, should be used to fill the trench, which should then be covered with a metal grating. Where oil can escape into an adjoining building, a trench should be installed at the threshold of each doorway, being either filled with pebbles or connected by a duct to an external sump.

For oil-filled switchgear with a capacity exceeding 100 litres, the escaping oil may be contained by providing a common fall pipe under the individual components of the switchgear to a sump outside the substation.

One of the most common methods of oil catchment for transformers is by the provision of dwarf walls. These should be of sufficient height to contain the full volume of oil inside the transformer; it is also desirable to fill the area enclosed by the walls with graded pebbles.

Instead of dwarf walls, the oil may be drained away by standing the transformer on rolled-steel joists over an internal sump filled with graded pebbles. Possibly a better method is to locate the sump outside the transformer cubicle, requiring only the addition of a fall pipe.

Oil may also be prevented from spreading from transformer chambers to switchrooms by terminating cable ducts between 75 cm and 100 mm above floor level and sealing the ends by packing asbestos or grouting hard-wood blocks around the cables. This would also assist in controlling the movement of vermin inside the ducts and helps to seal off obnoxious and explosive gases.

Where lead-sheathed cables are contained solely within the station, no further protection of the sheath is necessary, but cables with a protective covering of bituminized jute should have any exposed parts half-lapped with asbestos tape, or the serving should be removed. It is also advisable to seal off ducts and trenches containing the medium-voltage cables, particularly where they enter the substation or run between adjacent chambers in the substation itself. This helps to confine a cable fire and prevent it from spreading.

In all substations, it is also necessary to install fire-fighting equipment. For most stations, it is usually sufficient to place manually-operated fire-fighting equipment and appliances in readily accessible positions. In important substations, automatic fire-protection equipment is more often preferred. This is provided by installing cylinders of CO_2 gas, which operate when the temperature inside the station, rising above a pre-determined value, melts fusible links. When any of the links melt, the station is immediately flooded with carbon-dioxide gas, which quickly blankets the flames, thereby cutting off the oxygen supply to the flames.

7. *Compare the advantages and disadvantages of fuse-switches, oil-filled circuit-breakers and air circuit-breakers for the control gear of an industrial installation.*

The advantages of the fuse-switch (Fig. 7(a)) are:

(1) It is simple to control and therefore easy to maintain.

(2) It occupies little space, and can be easily mounted on busbar panels, although a far cleaner arrangement is produced by mounting it as an integral part of a flush-fronted switchboard.

(3) It is more economical to install than breakers.

(4) Its great advantage is its ability to clear faults of the utmost severity (up to 50 MVA) with great rapidity and precision, far faster than the circuit-breaker for short-circuit and earth-faults of high magnitude. As many engineers consider that the protective equipment has to be designed for the clearance of such faults, the fuse-switch with h.b.c. fuses is often preferred.

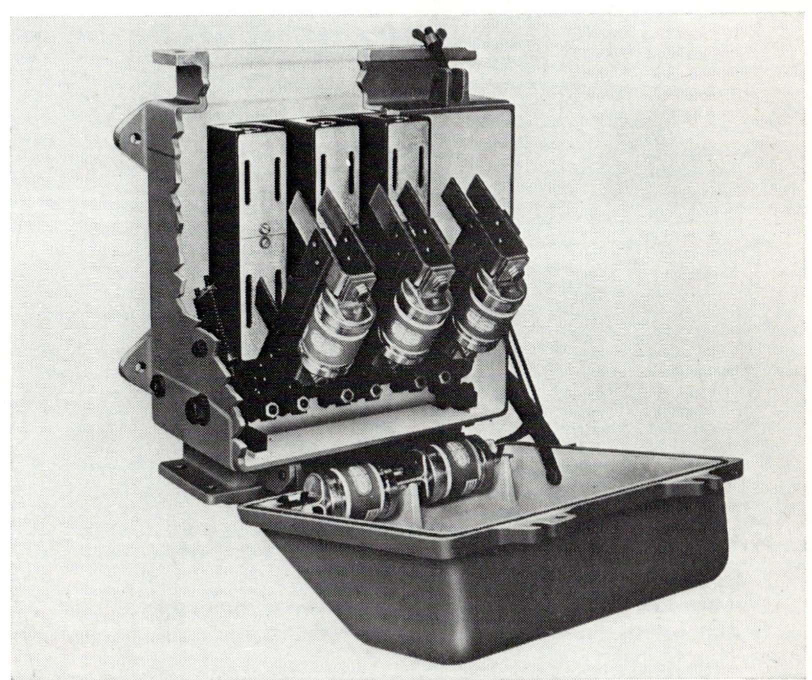

FIG. 7 *Construction of fuse-switch, air- and oil-filled circuit-breakers.*
(a) *Triple pole-and-neutral fuse-switch.* (*Courtesy English Electric Ltd.*)

The disadvantages are:

(1) There is no undervoltage or moderate overcurrent protection.

(2) The time required to replace a fuse is far greater than that necessary to close a circuit-breaker.

It should be noted that h.b.c. fuses can get very hot, which makes them difficult to handle immediately they have been ruptured.

Although fuse-switches are relatively cheap to install, the replacement of large h.b.c. fuses can be expensive. It is necessary, therefore, to ensure that any faults are cleared before fitting a replacement fuse.

The advantages of the oil-filled circuit breaker (Fig. 7(b)) are:

FIG. 7(b) *The fixed and moving contact assembly of a GEO* 400 A *circuit-breaker.* (*Courtesy George Ellison Ltd.*)

(1) It provides undervoltage and moderate excess-current protection.

(2) Although the breaker may be manually operated, it may also be opened or closed electrically by remote switches.

The disadvantages are:

(1) The risk of fire from escaping oil.

(2) The fault clearance is not so fast when severe faults occur as with the fuse-switch with h.b.c. fuses. Hence, where circuit-breakers are installed, many engineers prefer to back them up with h.b.c. fuses for heavy fault protection.

(3) The oil-filled breaker requires more maintenance than either the fuse-switch or the air circuit-breaker.

The advantages of the air circuit-breaker (Fig. 7(c)) are:

(1) It gives undervoltage and moderate excess-current protection.

(2) There is no need to make provision for the catchment of escaping oil.

(3) It occupies far less space than the oil-filled breaker and is very suitable for two- or three-tier arrangements in flush-mounted switchboards.

(4) The air circuit-breaker requires less maintenance than the oil-filled breaker but more than the fuse-switch, although the maintenance work is facilitated by the easy withdrawal of the breaker where it is contained in a withdrawable unit.

(5) As with the oil-filled breaker, the breaker may be operated manually or by remote switches.

FIG. 7(c) *Metal clad medium voltage air circuit-breaker, horizontal draw-out type.* (*Courtesy English Electric Ltd.*)

For medium-voltage supplies, the disadvantages of the air circuit-breaker are few, but one is that where fire hazards are present, such as in coal mines, refineries, chemical plants, etc., the installation of air circuit-breakers is inadmissible. Where, however, the arc itself does not constitute a hazard, the air circuit-breaker would generally be preferred now to the oil-filled breaker but possibly not to the fuse switch, especially for relatively low-rated circuits.

8. *Describe the construction of an oil-filled circuit-breaker and state what maintenance duties are necessary in order to maintain the breaker in sound working condition.*

The main parts of the breaker are:

(1) The fixed contacts mounted on insulated supports, the number and size depending on the value of the continuous current to be carried.

(2) The moving contacts, normally of silver-plated copper, are mounted separately on an insulated bar. When the breaker is closed, the moving elements make contact with the fixed elements and are held in position by mechanical pressure. When considering the heating effect at the surface of the contacts, the contact pressure is as important as the effective area of the contact surface.

(3) Arcing contacts are normally fitted in large circuit-breakers, consisting of small copper strips close to the main contacts. The purpose of these is to break the circuit before the main contacts, so preventing excessive wear on the latter.

(4) The operating mechanism for opening and closing the switch contacts. The mechanism is designed as a system of toggles and links which may be operated manually, mechanically or electrically.

(5) The framework which carries the breaker and the external connections, etc., and the tank containing suitable linings and contact barriers.

Periodically, the tank should be lowered and the contacts examined. If any roughness is observed, this should be removed and the contacts adjusted to give a firm and even pressure over their surfaces. The main contacts are usually silvered, and should the silver show signs of excessive wear, the contacts should be replaced immediately.

Arc contacts which, for oil-filled equipment would be of copper, possibly silver-plated, should be replaced when worn and before the main contacts take over the duty of making and breaking the circuit. It is important to ensure that all poles make simultaneously on both the arcing and main contacts, respectively.

The oil should also be examined for deterioration in quality, possibly by comparing the colour with that of a sample of fresh oil. One purpose of the oil is to assist in cooling the contacts. The oil, however, tends to become carbonised should the contacts overheat, and where this occurs, the contacts should be cleaned, the tank cleaned and re-filled with oil of the correct grade and to the right level.

In addition to the oil and contacts, the following maintenance work should also be undertaken:

(1) Dust and dirt must be carefully removed.

(2) All bearings and moving parts must be lubricated.

(3) All connections should be examined to ensure they are in good condition.

(4) Worn parts must be replaced.

(5) Time lags should be inspected to ensure that the right grade of oil has been used and that the dash-pot is filled to the correct level.

(6) Trips should be tested to ensure that they operate freely and consistently.

(7) All insulated parts should be cleaned and examined for any deterioration or signs of breakdown.

9. *What are the points to be considered when installing transformers in substations?*

The points to be considered are:

(1) The temperature rise anywhere inside the transformer must not exceed the permissible limit and, therefore, the layout should be planned so as to allow sufficient room between individual units and between units and adjacent walls. The actual separation, however, also depends on the capacity of the transformer and the type of cooling system adopted.

(2) Precautions must be taken against fire and escaping oil (see question **8**).

(3) In addition to adequate space for ventilation, there must be unobstructed access to drain cocks, so that samples of oil might easily be removed for testing purposes, and to thermometer pockets so that temperature recordings of the oil may be taken at regular intervals.

(4) The foundations must be satisfactory. In general, transformers stand on concrete floors and holding-down bolts are unnecessary if the floor is level. Other structures are permissible, provided they will carry the concentrated weight of the transformer. It is essential, however, to select a location not subject to appreciable vibration which might prove harmful if transmitted to the windings.

(5) Sufficient headroom must be allowed for lifting transformer cores and windings out of the tank, and it is advisable when the substation is being constructed to install a rolled steel joist over the transformers for anchoring lifting tackle. If it is not possible to remove the core and windings without dismantling the transformer, it is advantageous to fit swivelling rollers to the underside of the transformer so that it can easily be removed for maintenance purposes.

10. *Describe what is meant by discrimination and state how it is applied in small and large installations.*

Discrimination means that where a circuit or part of an installation is faulty, only the device protecting that circuit or that part of the installation must operate, and, therefore, no other healthy circuit has its supply interrupted. Thus, in Fig. 8 should a fault occur at X, only the fuse or circuit-breaker controlling the section should be disconnected.

In small installations, where semi-enclosed fuses are frequently used, accurate discrimination is difficult because of the manner in which the fuse deteriorates on load. For this reason, it is recommended that the rating of the sub-circuit fuse

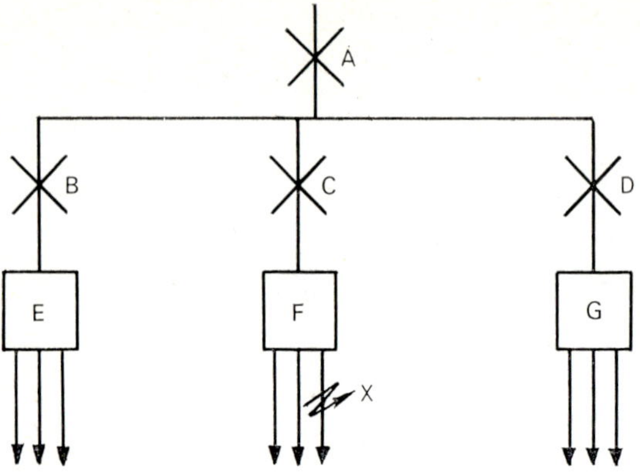

FIG. 8 *Illustrating discrimination.*

should not exceed 50 per cent of the rating of the main fuses, and the same make and design of fuse should be used throughout the installation.

Where cartridge fuses are used for the main circuit and semi-enclosed for the sub-circuits, it is virtually impossible to attain discrimination where heavy short-circuit faults are liable to occur. In long sub-circuit runs where the resistance of the cables is relatively high and the prospective short-circuit current is limited, more accurate discrimination is possible, and in such situations the rating of the sub-circuit fuse should never exceed 50 per cent and preferably 25 per cent of the main fuse.

Where cartridge fuses are used for both the main fuses and sub-circuit fuses which is frequently found in large installations, the main and sub-circuit fuses should preferably be of the same type, make and design. In general, discrimination is obtained if the rating of the sub-circuit fuses do not exceed 50 per cent of the rating of the main fuses. Where very large prospective currents are possible, and where better discrimination is required, a wider margin between the main and sub-circuit fuses is desirable.

In large installations cartridge fuses are often used to back up circuit-breakers. In such situations the circuit-breaker gives protection against sustained moderate overloads while the fuse gives protection against heavy short-circuit faults. Where cartridge fuses are used for the main circuits, and circuit-breakers for the sub-circuits, again accurate discrimination is not always possible, because the fuse is liable to operate before the circuit-breakers under heavy short-circuit conditions, although under sustained moderate overload conditions the circuit-breaker would operate before the fuse.

In large installations, circuit-breakers are often used for both main circuits and sub-circuits. Accurate discrimination can be provided by arranging for the sub-circuit breakers to operate at a lower overload current setting and a shorter time-lag than that of the main circuit-breaker. For this reason the excess-current setting on circuit-breakers is always made adjustable.

2

Consumer's circuits

Lighting circuits—13 A *socket-outlet circuits—bell circuits—fire-alarm circuits.*

11. *Describe how a light may be individually controlled from three different switch positions.*

There are three different types of intermediate switch and it is important to remember that where a faulty intermediate switch has to be replaced, if the new switch is of a different type, then the wiring to the switch has to be altered, otherwise the circuit will not operate satisfactorily. Fig. 9 illustrates the three different types of intermediate and two-way switching.

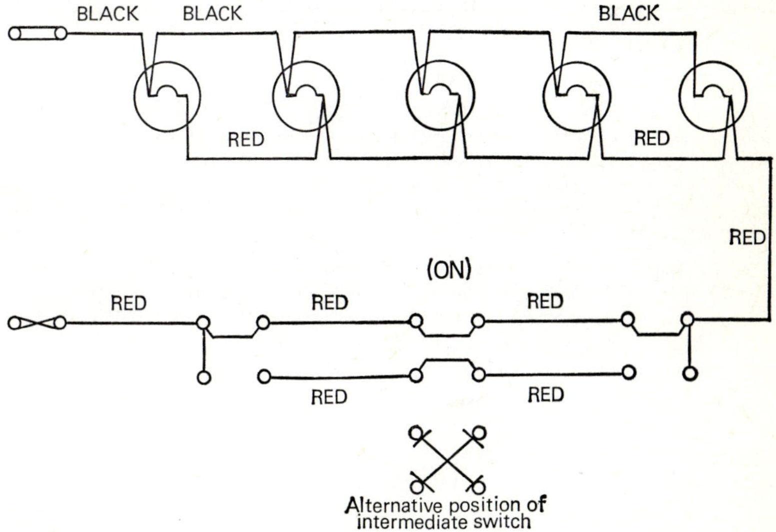

FIG. 9(a) *One light controlled individually from three positions.*

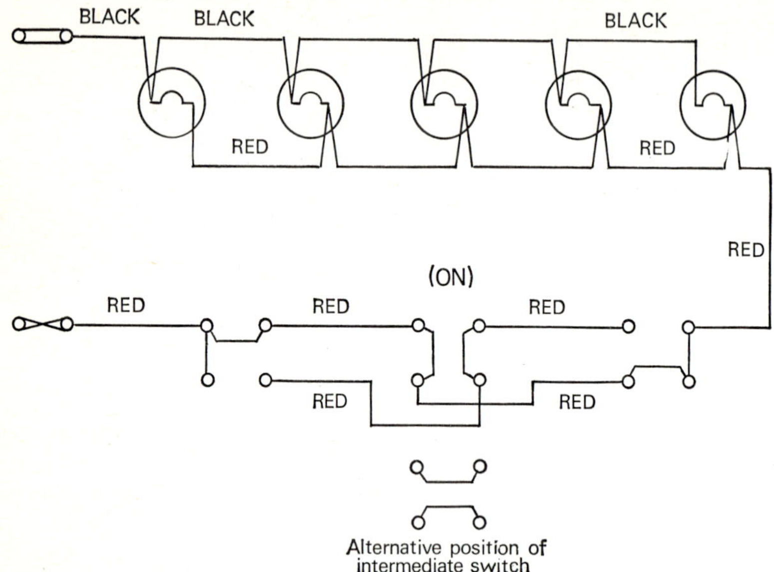

BLACK
BLACK
BLACK
RED
RED
RED
(ON)
RED
RED
RED
RED
RED
Alternative position of intermediate switch

BLACK
BLACK
BLACK
RED
RED
RED
(ON)
RED
RED
RED
RED
RED
Alternative position of intermediate switch

FIG. 9(b)

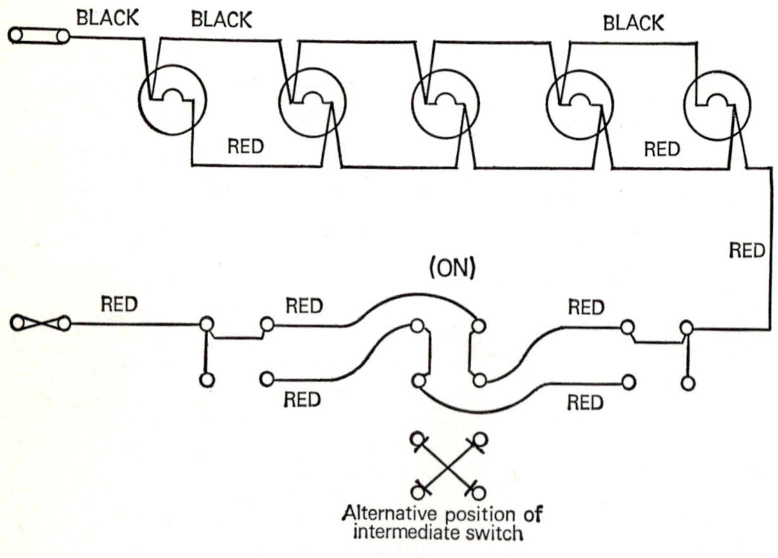

FIG. 9(c)

12. *Show by line diagrams all the permissible circuit arrangements for* 13 A *socket-outlets, and include any special conditions. Where must caution be exercised in installing ring-main circuits of* 13 A *socket-outlets?*

Fig. 10 illustrates all the circuit arrangements for 13 A socket-outlets. Note that circuit (d) merely shows one arrangement for a ring circuit as the number of socket-outlets permissible is unlimited, subject to there being at least one circuit for every 100 m² of floor area. It also shows that the number of spurs may be the same as the number of socket-outlets on the ring, but not more.

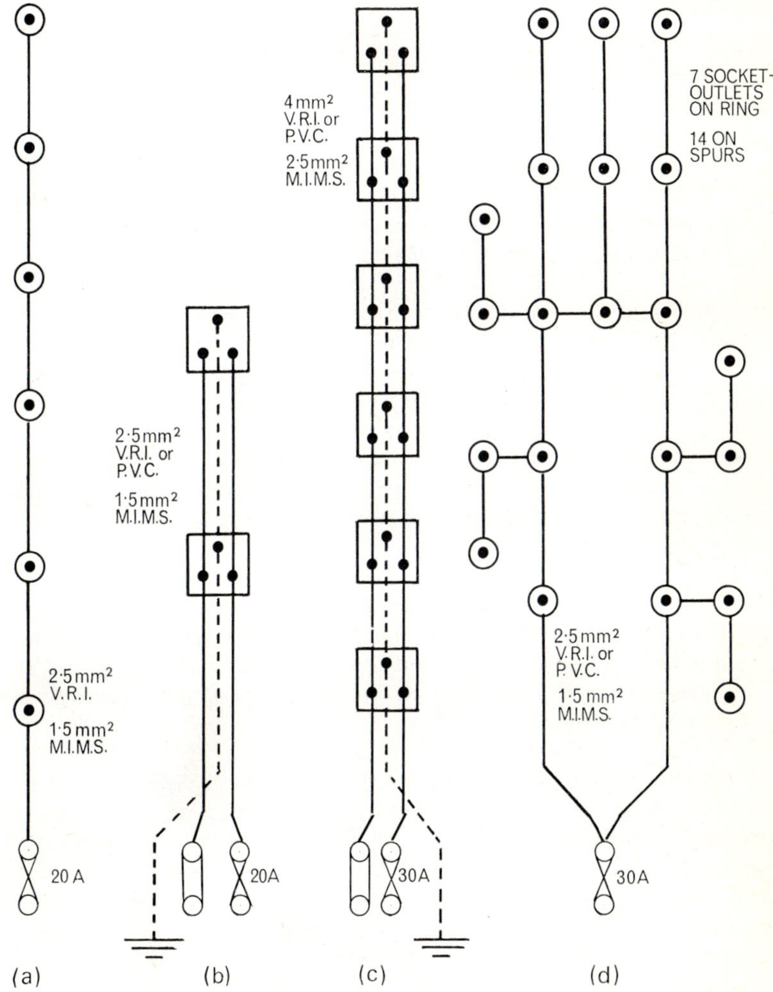

FIG. 10 13 A *socket-outlet circuits.*

Where a ring main contains ten 13 A socket-outlets, the possible load that could be connected to the circuit is 130 A, but, because of the fuse rating, the actual permissible load must not exceed 30 A. The diversity factor is therefore

$$\frac{30 \times 100}{130} = 23 \text{ per cent.}$$

It is clear, therefore, that only a small proportion of the socket-outlets can be fully loaded at the same time, and where a large number of the sockets are liable to be fully loaded, such as in a commercial installation where the socket-outlets are required to serve heating appliances, then such a circuit would be overloaded and would be inadmissible.

13. *Describe what is meant by restricted signalling, and state how it is applied to an open-circuit fire alarm system in a hospital.*

The hospital consists of three buildings. One alarm and four pushes are required in each building, the bells to ring only in the building in which the call is originated. Essential bells are required at various points in the installation and a supervisory sounder and indicator are required at the fire post.

Restricted signalling is installed where the ringing of bells in places other than the one where the call is originated, would cause unnecessary inconvenience and possible distress. By the installation of relays at the required points in the installation, only the essential bells and the bells associated with the operated call-point ring.

In the circuit illustrated (Fig. 11), the main bells are switched off by means of a manually operated series of two-way switches which when turned in the other direction, closes the circuit for the supervisory sounder.

14. *With respect to fire alarms, what is meant by a closed circuit and how is it applied to the following closed circuit in a hospital consisting of three buildings? One alarm bell and four pushes are required in each building, the bells to ring only in the building in which the call is originated. One main bell and section indicator are required at the fire post.*

In the closed-circuit system the call-points have normally-closed contacts, and a relay is incorporated which is constantly energised while the circuit is connected to the supply. When the call-point is operated, the relay coil is de-energised releasing an armature which closes two contacts in the main bell circuit. This rings the bell associated with that push and the main bell. At the same time, the indicator flag associated with the building in which the call is originated falls, so giving a visible indication of where the call-point has been operated. When the glass in the call-point is replaced, the relay coil is again energised, so causing the bells to stop ringing.

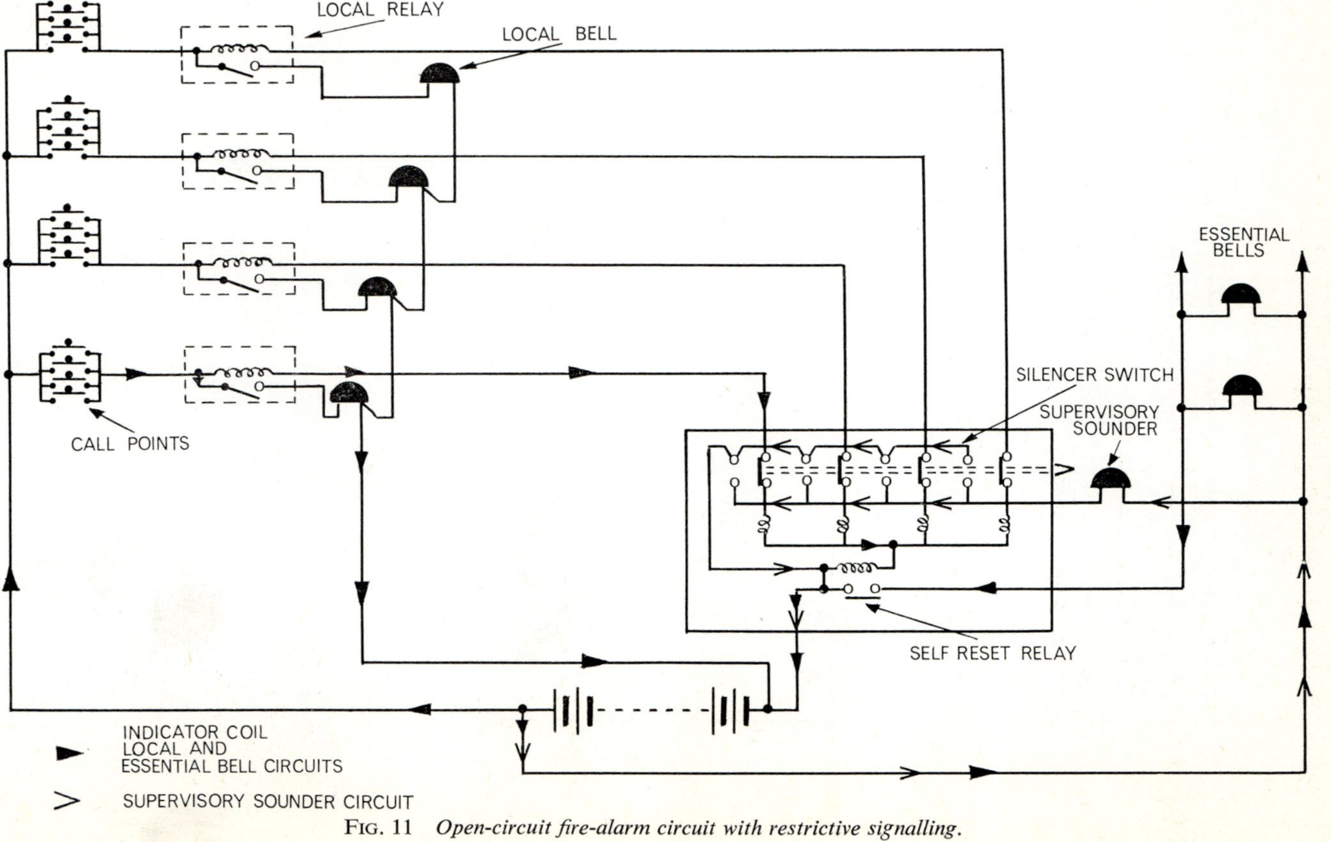

FIG. 11 *Open-circuit fire-alarm circuit with restrictive signalling.*

Fig. 12 illustrates the closed circuit fire alarm system.

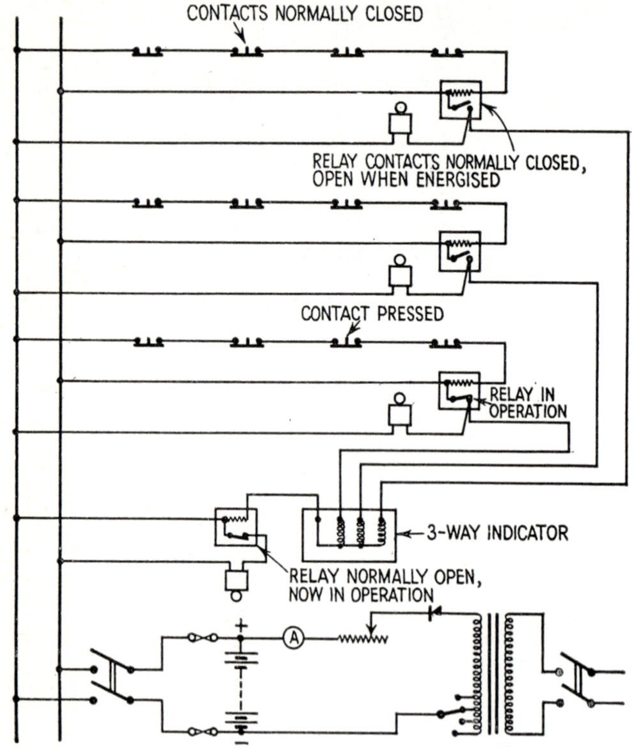

FIG. 12 *Closed-circuit fire-alarm circuit with restrictive signalling.*

15. *Describe the purpose of an automatic detector in a fire-alarm circuit and its construction and operation.*

Automatic detectors are of particular importance in premises or parts of premises where the fire risk is great, and where continuous protection is necessary in those parts which are left unattended for long periods. The installation of automatic detectors and manually operated call-points provide virtually a complete fire protection.

Whichever type of automatic detector is installed, the principle is similar, in that, when an abnormal temperature occurs, the detector operates, so opening or closing contacts in the alarm circuit, depending upon the type of circuit installed.

The most common type of automatic detector (Fig. 13) contains a mercury switch mounted on a bimetallic strip. An abnormal temperature rise causes the strip to flex, the operation releasing a catch which allows the lower part of the detector to tilt. This allows the mercury switch to make or break, as required, to operating the alarm.

FIG. 13 *Construction of an automatic fire detector.* (*Courtesy Gent & Co. Ltd.*)

Such a detector operates at a nominal operating temperature of approximately 65° C, but where there is a slow temperature rise, the operating temperature is about 59° C, and will operate within 2·5 minutes when the temperature rises at a rate of 22° C per minute from normal room temperature.

The components parts are corrosion proof and the detector is designed for attachment to standard conduit boxes. Each detector is capable of protecting an area 6·1 m square, so that by dividing the total area by 37 gives the required number of detectors for complete protection.

16. *Describe the purpose of an alarm stop/reset relay in a fire alarm circuit. Describe also its construction and operation.*

The purpose of the alarm stop/reset relay is to disconnect the main bells and transfer the call to a supervisory sounder at the fire post, which remains ringing until the glass in the operated call-point has been replaced.

The relay comprises an electromagnet and a two-way switch, the connections of which are shown for a closed circuit system in Fig. 14. When a call point is pressed the main bells start to ring through one connection of the two-way switch in the relay. A manual device operated from outside the relay, is pressed which disconnects the main bell and transfers the call to the supervisory buzzer. Simultaneously, the relay coil is energised, so that the supervisory sounder remains ringing. When the call-point glass is replaced, the circuit to the supervisory sounder and relay call is opened so that the circuit returns to normal.

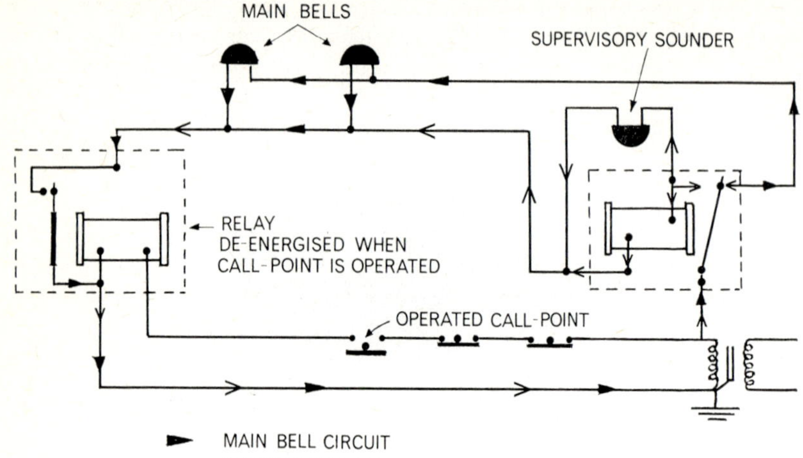

FIG. 14 *Wiring diagram of an alarm stop/reset relay.*

3

Cables

Rating factors—selection of cables—sealing—corrosion in sheath—joining plastic mains cables and aluminium-cored cables.

17. *Describe the types of cable most commonly used in installation work and state the situations in which they are most likely to be installed.*

The following types are recognised in the Regulations as being suitable for low- or medium-voltage wiring:

(1) *P.V.C. insulated cables* (Fig. 15(a)). Polyvinyl chloride is the general term for a class of thermoplastic materials based on polymers of vinyl chloride modified with plasticizers and other ingredients. Conductor sizes vary from 1 mm^2, the smallest size, to 630 mm^2, the largest size. The smaller sizes are used for wiring systems and the larger sizes for distribution systems.

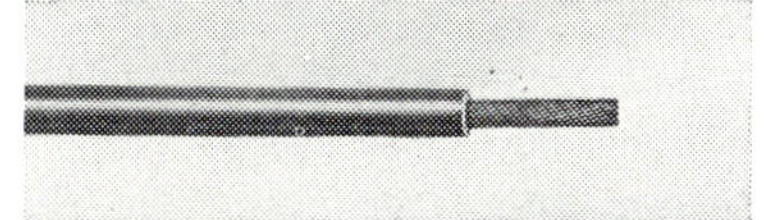

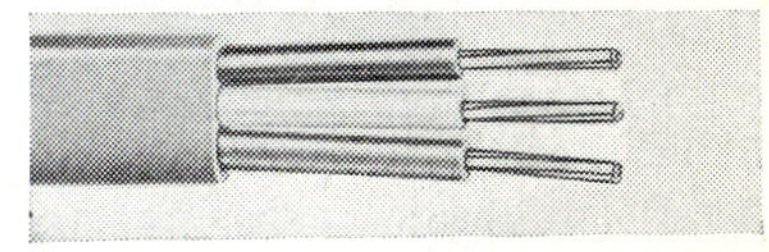

FIG. 15 *Illustrating types of cables.*
(a) *Three-core, p.v.c. insulated and p.v.c. sheathed cable and single-core p.v.c. insulated cable. (Courtesy Johnson & Phillips Ltd.)*

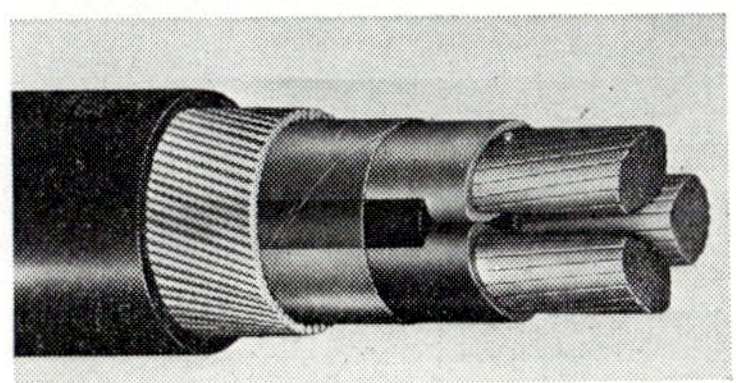

FIG. 15(b) *Armoured three-core, P.V.C. insulated cable with copper conductors. (Courtesy BICC Ltd.)*

The insulation is extruded over the conductors, the thickness depending on the grade of the cable. There are two grades for normal installation work. 250/440 V and 660/1100 V. For conduit systems, the p.v.c.-insulated cable is satisfactory, but for other wiring systems, two or more insulated cores, possibly with a bare earth continuity conductor, are contained in a plastic sheath, which provides extra mechanical protection. Large mains cables (Fig. 15(b)) would also possibly have a wire armouring for additional mechanical protection which, in turn, would be protected, where necessary, by a plastic oversheath against corrosion (Fig. 15c).

P.V.C. flexible cords (Fig. 15(c)). These may be obtained in the twisted twin, non-sheathed type which is suitable for pendants; the parallel twin type which is used for wiring lighting fittings and appliances where the cord is not subject to abrasion and where earthing is unnecessary; and the three-core circular type which is used where an earth continuity conductor is required. It is generally preferred in industrial installations where oil and water are present.

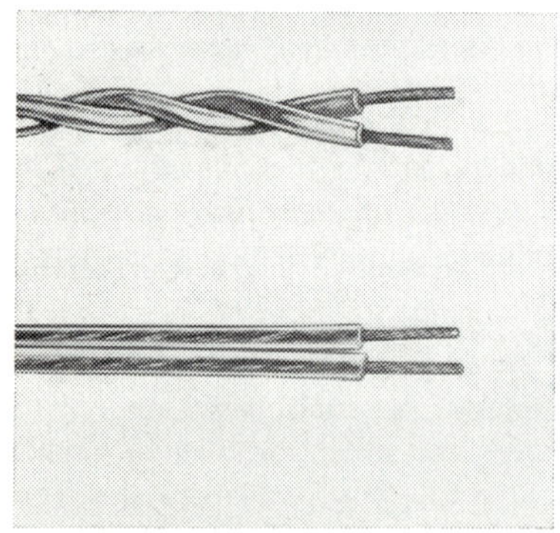

FIG. 15 (c) *P.V.C.-insulated parallel twin flexible core (Courtesy Johnson & Phillips Ltd.)*

(2) *Vulcanised-rubber-insulated cables.* For most installation work, p.v.c. insulated cables have superseded vulcanised-rubber-insulated cables, but any of the cables insulated with p.v.c. may be obtained with rubber insulation. The rubber is vulcanised to give it additional strength. As sulphur is used during the vulcanising process, the conductors are tinned to prevent corrosion. An identification tape is then wrapped round the insulation and the cable is finally finished with a waxed cotton braiding.

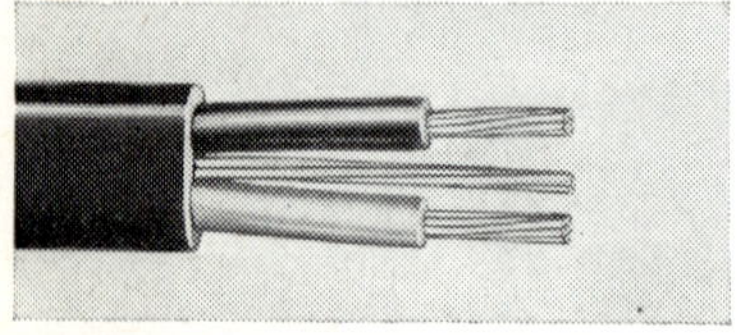

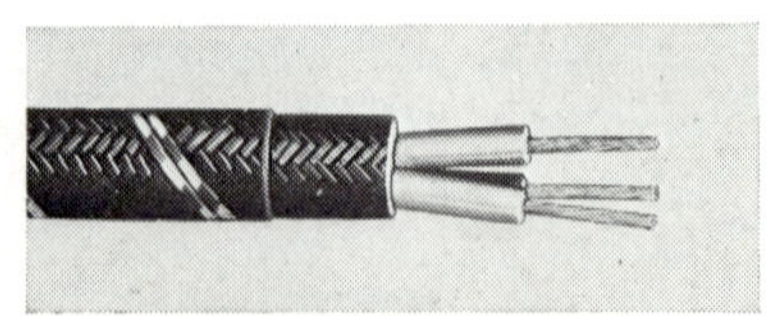

FIG. 15 (d) *Three-core rubber-insulated braided domestic cord and T.R.S. twin and earthed cable. (Courtesy Johnson & Phillips Ltd.)*

For twin conductor wiring systems, the insulated core(s) is further protected by a tough rubber sheath (Fig. 15(d)).

One disadvantage of this type of cable is that it cannot be installed in direct sunlight, but it has one advantage in that it can be installed in temperatures far lower than those in which p.v.c.-insulated cables can be installed.

(3) *Paper-insulated lead-sheathed or aluminium-sheathed cables* (Fig. 15(e)). In this type of cable, the conductors, which may be of copper or aluminium are covered with layers of high-grade paper approximately 5 mm wide, the number of layers depending on the voltage of the system in which the cable has to operate. The cable is normally impregnated with oil-resin compound, although for vertical runs or steep gradients special compounds may be applied. The need for special compounds in long vertical runs is because there is a tendency for oil to migrate through the normal compound into accessories and boxes. Furthermore, there is a possibility that the normal compound may build up sufficient pressure to burst the sheath of the cable where it is installed vertically.

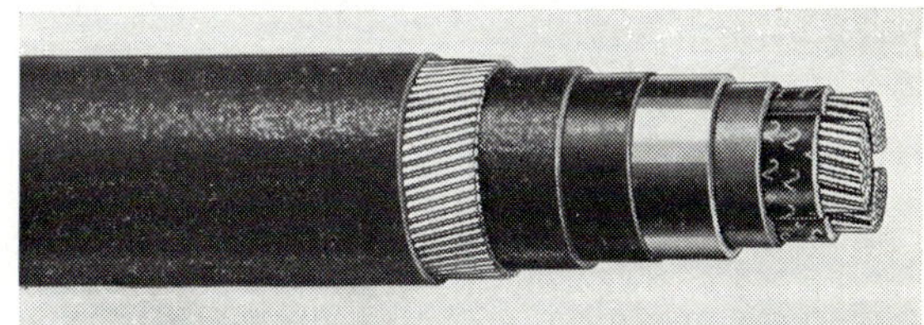

FIG. 15 (e) 195 *mm*^{2} *three-core* 11 *kV cable, paper insulated, lead sheathed, steel-wire armoured and served.* (*Courtesy Johnson & Phillips Ltd.*)

The cable, encased in its lead sheath, is only suitable for certain locations where mechanical protection is unnecessary. Normally, therefore, a single-wire-armoured, double-wire-armoured or double steel-tape-armoured cable is installed. Since it provides a better protection against glancing blows, the double-steel-taped armoured cable is usually installed for underground services, but for services above the ground or where longitudinal stresses are likely to be imposed on the cable, the wire-armoured cables would be preferred.

For normal industrial services, the p.v.c. armoured cable has superseded the paper-insulated cable, but for high-voltage systems, the paper-insulated cable is preferred. A further advantage compared with cables with other types of insulation is its high current rating.

Its disadvantage is that the paper is hygroscopic and, therefore, special sealing precautions have to be taken at terminations. This calls for highly skilled jointing techniques, especially for very high voltage cables.

(4) *Varnished-cambric-insulated cables* These are similar to paper-insulated lead- or aluminium-sheathed cables except that the insulation consists of a textile impregnated with insulating varnish. The cable is often preferred to the paper-insulated cable where it is important there should be no seepage of compound. Moreover, for indoor installations sealing boxes are not essential provided the ends are protected by taping.

(5) *Mineral-insulated metal-sheathed cables* (Fig. (15f)). In this type of cable solid copper or aluminium conductors are embedded in an insulant of highly compressed magnesium oxide, the conductors and insulant being contained in a sheath which may be of the same metal as the conductor, or in the latest type of cable, copper conductors enclosed in an aluminium sheath. Although the conductors are located very close to one another, their relative positions are not disturbed by bending or twisting the cable. Indeed, the cable is capable of withstanding severe mechanical stresses without breakdown between conductors.

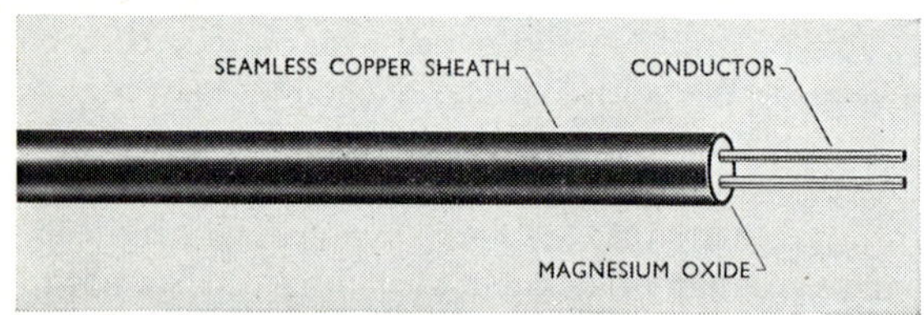

FIG. 15(f) *Twin m.i.m.s. cable.* (*Courtesy BICC Ltd.*)

Some of the advantages of this type of cable are:

(*a*) It is very malleable, special bending levers being necessary only for the larger sizes.

(*b*) It is fire-proof, non-fire raising and virtually ageless.

(*c*) It is less obtrusive than other types of similar rating.

(*d*) The current rating is higher than that of other types of similar size.

(*e*) The temperature at which the cable can be operated is limited only by the type of seal and the possible oxidation of the sheath.

(*f*) The cable is mechanically strong and resistant to heavy blows and violent twisting.

The disadvantages are:

(*a*) The insulant is hygroscopic which necessitates special sealing techniques. The type of seal has to be selected to suit the conditions in which the cable is to be installed.

(*b*) It is not so economical to install as other systems, for example t.r.s.-and p.v.c.-sheathed wiring systems.

In general, mineral-insulated cables are very suitable in flammable situations or where the ambient temperature is high; by using special terminations the cable can operate in ambient temperatures up to 145°C. The cables are immune to the effects of moisture and are therefore suitable for installing in damp and exterior situations. By enclosing the copper sheath in a plastic covering and using plastic shrouds at terminations, the cable can be used in many corrosive situations. Because only two terminations are normally needed, and because of its simplicity of installation, the cable has a wide application for mains and sub-mains. As the cable is immune to condensation, it is preferred where such problems exist.

The necessity for differentiating between the two types of protection is that p.v.c.-insulated cables and cables insulated with synthetic rubber are subject to damage or deterioration when they are subjected, even for relatively short periods, to temperatures appreciably higher than those permissible under continuous operation. The current rating of such cables is determined therefore not only by the temperatures which might be obtained during continuous running conditions, but also by the temperature obtainable during excess-current conditions.

To obviate the necessity of duplicating tables for cables with both types of protection, for cables with loads up to 100 A and insulated with p.v.c. or synthetic rubber, the current ratings and voltage drop per ampere per metre are based upon the proviso that the circuit is designed to operate with coarse excess-current protection, and should the circuit be protected by a close excess-current protective device, the values may be up-rated by applying a rating factor of 1·33 for p.v.c.-insulated cables and 1·22 for butyl or silicone rubber-insulated cables.

Where the current rating exceeds 100 A, the circuit would normally be protected by h.b.c. fuses, and therefore the tables where the cable size exceeds 35 mm^2, are based upon close excess-current protection. Should coarse excess-current protection be provided, then a factor of 0·75 for p.v.c. cables and 0·82 for butyl-insulated cables must be applied.

20. *Describe the factors to be taken into consideration when selecting cables for particular loads and types of situation.*

Once the current to be carried has been calculated, it is necessary that the cable selected should be able to carry the current without producing any excessive temperature rise for the conditions under which it is installed. Where the run is moderately long, however, the voltage drop in the cables becomes the most important consideration. Thus when a 2·5 mm^2 p.v.c.-insulated twin cable is carrying 21 A, the voltage drop in the cables is 8·4 V for a run of 30 m. Even if the full voltage drop (6 V) was allowed in the cable, it would still be above the permissible value, and in industrial situations, runs exceeding 33 m are exceedingly common. To keep the voltage drop to 6 V, either the current would have to be reduced to 15 A or the length of run to 24 m. Therefore, if the length of run and the current had to be maintained, then a larger size of cable would have to be installed. It is a well known axiom in installation work that for long runs, oversize cables are normally necessary.

The type of cable to be installed also depends upon the situation in which it is to operate. Thus:

P.V.C.-sheathed cables must not be installed where they are open to mechanical damage, or they must be protected against such damage.

Aluminium-sheathed cables must be protected against corrosion where installed in concrete ducts or underground pipes and wherever there is a possibility of corrosion due to damp or corrosive elements.

Mineral-insulated cables must also be protected against corrosive elements (such as sulphur), where present, by a covering of p.v.c. and using plastic shrouds at terminations.

Cables must also be selected to suit the ambient temperature in which they have to operate, that is p.v.c. cables must not operate at temperatures in excess of 70°C.

P.V.C.- or paper-insulated cables should not be installed during periods when the temperature is below 0°C.

Flexible cords and cables must also be selected to suit the situation in which they are to be installed. New tables (Tables 21 and 22) added to the I.E.E. Regulations specifically relate to high-temperature flexible cords and cables. Sheathed flexible cords and cables must be installed in situations where mechanical damage is possible, and, in particularly onerous situations, they must also be metal sheathed.

21. *Describe why it is essential to seal paper-insulated cables and mineral-insulated cables and state the types of seal available for the latter type of cable.*

The insulation of paper-insulated cables consists of layers of impregnated paper while the insulation of mineral-insulated cables consists of a magnesium oxide powder. As both types of insulation are susceptible to damage through the ingress of moisture, special seals or sealing boxes must be provided to prevent such entry.

The types of seal available for mineral-insulated cables are:

(1) The cold seal which is used for general purpose work and where the ambient temperature does not exceed 75°C. General purpose compound and Neoprene (p.c.p.) or p.v.c. sleeving is used for the insulation.

(2) The medium temperature seal which is used where the ambient temperature does not exceed 100°C. For this type of seal, special plastic compound, Silastic rubber sleeving, silicone-bonded glass caps and porcelain anchoring wedges are required.

(3) The oil-resistant seal which is similar to the cold seal except that a special compound is employed and must be applied when the cable and seal have been heated.

(4) The glazed seal termination which is used where the ambient temperature does not exceed 145°C. A special powder is melted into the pot and insulating beads and a ceramic cap must also be used.

22. *Describe the advantages and disadvantages of plastic mains cable where used in industrial installations.*

The advantages of the plastic mains cable are:

(1) It is relatively economical to install.

(2) One of its good electrical properties is its insulation resistance to water and certain chemicals.

(3) The migration of compound is minimised.

(4) The jointing of the cable is relatively simple and can generally be undertaken by competent electricians.

(5) Where aluminium conductors are incorporated, a given run of this cable is much lighter than the same run of cables of similar rating with copper conductors.

(6) As no lead sheath is required, plumbing of the sheath to accessories is not required.

The disadvantages of the plastic mains cable are:

(1) The protective devices are more critical in operation (see question **19**).

(2) Because of the absence of a lead sheath, it is sometimes difficult to obtain a sufficiently low resistance of the earth-continuity conductor to satisfy the Regulations. In such cases, it is necessary to incorporate a further core to supplement the earth-continuity conductor.

(3) The cable should not be installed in conditions of extreme temperature, At low temperatures, the insulation tends to crack, and at high temperature, the insulation tends to soften. Also, continuous exposure of p.v.c. compounds to temperatures above 115°C may cause corrosive products to form which can attack conductors and other metalwork.

23. *Why are single-core steel-wire or steel-taped armoured cables unsuitable for a.c. supplies? Where metal-sheathed single-cored cables have to be connected in parallel, what effect is caused and what precautions should be taken against such effect?*

Where single-core cables are surrounded by a metal sheath, eddy currents are induced in the sheath when the cable is connected to an a.c. supply, leading to undesirable heat and power losses in the sheath. The effect is even more pronounced where the cable is further protected by a steel armouring.

Where large currents have to be carried, it is often necessary to install single-core cables in parallel. On a.c. supplies, therefore, this effect can produce circulating currents in the sheaths if the sheaths are bonded together, or standing voltages if the cables are left unbonded.

If the sheaths are bonded but the current is below the full-load value, the additional heating caused by the circulating currents is not detrimental to the cable. If the cable sheaths are not bonded and standing voltages are present, there is a possibility of discharge between unprotected sheaths, which might cause one of them to be punctured.

The circulating currents or standing voltages depend upon the current in the cable, the dimension of the cable and the spacing. Although bonding is usually adopted, there is no definite rule as to whether sheaths should be bonded or not.

The cables are usually only left unbo ey are required to carry their full load, and where there is little risk anding voltages being induced in the sheaths.

If the latter method is adopted e to use plastic-covered cables and to fit strong insulation boar ol at terminations.

Where ferrous plates are used currents being induced in them, slots should be made between Alternatively, the plates may be constructed of non-ferrous met s.

24. *Describe how corrosion n metal-sheathed cables and conduits and state what precautions sh to minimise such corrosion.*

One type of corrosion stalling bare steel armouring and steel conduit in damp situa ture in the atmosphere oxidising any unprotected armourin

A further type of c d where chemicals are also present in the atmosphere; a specif with which most electricians are familiar is the effect of sulphu hs of unprotected cables.

A third type is ytic action where two dissimilar metals are installed in a situ tentionally, an electrolyte is provided by the atmosphere.

Corrosion is a when metal sheaths or conduit are in contact with the following ma

Materials used in the construction of floors and dadoes, and which contain magnesium chloride.

Plastic undercoats contaminated with corrosive salts.

Lime, cement and plaster (for example, on unpainted walls).

Oak and other acidic woods.

Application of bitumen or bituminized paint before erection, or prevention of contact by separation with bitumen felt are recognised as effective precautions against corrosion, although where corrosion is likely to occur, all metal-sheathed cables should have a plastic covering.

In corrosive situations, special care must be taken in selecting clips and other fixings for aluminium-sheathed cables and aluminium conduits, suitable materials being as follows:

Aluminium.

Corrosion-resistant aluminium alloys.

Zinc alloys complying with B.S. 1004.

Iron or steel protected against rust by galvanizing, sheradizing, etc.

Contact between bare aluminium sheaths or aluminium conduits and any parts made of brass or other metals having a high copper content should be avoided in damp situations unless the parts are suitably plated. Where contact is unavoidable, the joint should be completely protected against the ingress of moisture. Wiped joints in aluminium-sheathed cables should always be protected against moisture by a suitable paint, by an impervious tape, or by embedding in bitumen.

Bare aluminium-sheathed cables and aluminium conduit are, however, quite satisfactory indoors in situations of normal humidity, free from chemical fumes or dust injurious to aluminium.

Protection, however, should always be given to aluminium cables in indoor situations where dampness is likely to occur after the cables have been installed, and for outdoor situations where corrosive elements are likely to be deposited on the sheath. Examples are:

Marine atmospheres where there is a possibility of chloride deposit.

Heavy soot deposition.

Injurious chemical fumes or dusts.

Damp walls or other absorbent materials.

Aluminium sheaths are particularly prone to what is termed *crevice attack*, where unprotected sheaths make direct contact with a damp porous surface which may be wood, paper, plaster, etc. The crevice is formed at the point of contact and corrosion may be rapid even in a normally unpolluted atmosphere or where the surface contains no chemical harmful to aluminium.

In addition to the foregoing precautions, where aluminium cables have to be supported they should be installed away from walls to avoid contact with absorbent materials.

Where cables pass through walls, floors, etc., they should be protected with two layers of p.v.c. tape extending to at least 80 mm either side of the wall, floor, etc. The hole should then be made good with cement or fire-resistant material. Cleats should preferably be of the all embracing type, and to avoid strain on cable joints, cleats must be installed reasonably close to terminal positions.

25. *Describe the making of a through joint in a plastic-armoured mains cable.*

The joint may be made in a cast-iron box or it may be contained in a plastic box (Fig. 16) subsequently filled with polyester resin. This filling material has an additional hardening agent, supplied in a separate tin. When both are poured into the box, heat is developed which causes them to combine and set into a solid homogeneous mass.

Since there is no lead sheath, to ensure continuity where plastic boxes are used, armour clamps with a bonding strip between them are essential.

The two cables are cut to the correct length, the plastic oversheath and armour are then removed just sufficiently for the armour to be held firmly by the armour clamp.

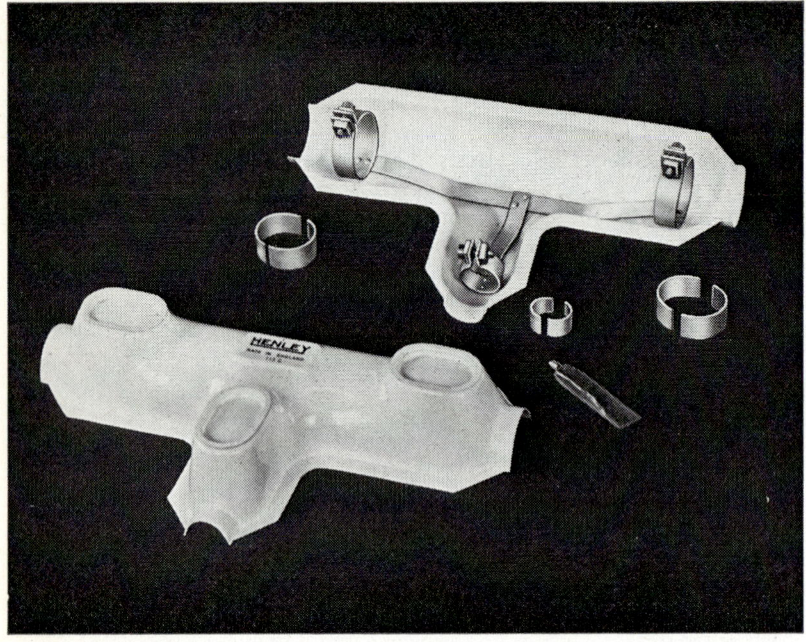

FIG. 16 *Plastic joint box. (Courtesy W. T. Henley's Telegraph Works & Co. Ltd.)*

The core insulation should be removed just sufficiently for the conductors to be enclosed by the terminal connection. If soldering is adopted, soldering ferrules should be used and the core insulation should be wrapped with nylon tape to avoid distortion of the insulation. Also, the heat should be applied for the shortest time possible. A more satisfactory method, however, is to employ compression (Fig. 17) or mechanical joints. If these are used, together with the cold pouring filler, all operations involving heat are eliminated.

The connections are then insulated with plastic tape to the required thickness and the joints are laid in the box which is then assembled and filled with compound. Care should be taken to ensure that the armour is firmly secured by the armour clamps and that both clamps are bonded together. The end of the plastic overcovering and the clamp may be wrapped with plastic tape and a good bituminous paint applied to prevent corrosion.

The box should be supported so that no strain is imposed upon the joint or conductors.

At terminations in dry situations, sealing of the cables is not required, and the fittings are confined to cable sockets (if required) and a suitable armour clamp or gland. These may be of the cone type for bonding the armour and may also have a conduit thread enabling it to be screwed into the necessary accessory, etc.

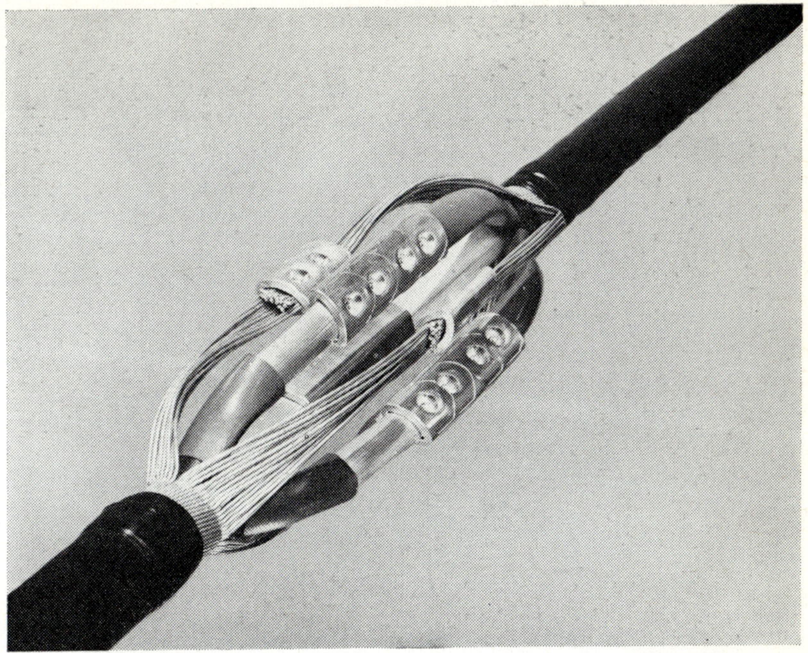

FIG. 17 *Through joint in plastic-armoured cable.* (*Courtesy BICC Ltd.*)

26. *Describe the making of a through joint in a paper-insulated, aluminium-cored and aluminium-sheathed cable.*

Any of the boxes used for jointing paper-insulated lead-sheathed cables may be used provided that protection is provided against corrosion. The difference lies in the techniques employed for tinning the sheaths. Both are highly skilled operations, but whereas the tinning of the lead sheath is quickly learned by the jointer, the tinning of the aluminium sheath seems to occasion more difficulty. This is because there is a tendency for the oxide to reform on aluminium immediately after cleaning, therefore a very active flux and special solder are necessary. The flux is a proprietary compound only active over a limited temperature range (300°C to 320°C), while the solder (referred to as Fryal metal) is a mixture of tin, lead and zinc.

One method of tinning is as follows:

The plumbing area is thoroughly degreased, using a rag soaked preferably in trichoethylene. Heat is immediately applied to the sheath and it is then possible to scratch brush a pool of solder over the required area with a wire brush until the whole area is completely tinned (Fig. 18). Some of the flux may be applied from time to time to assist the tinning process. Once the aluminium sheath has been tinned it may be plumbed to the lead sheath using ordinary plumbing metal.

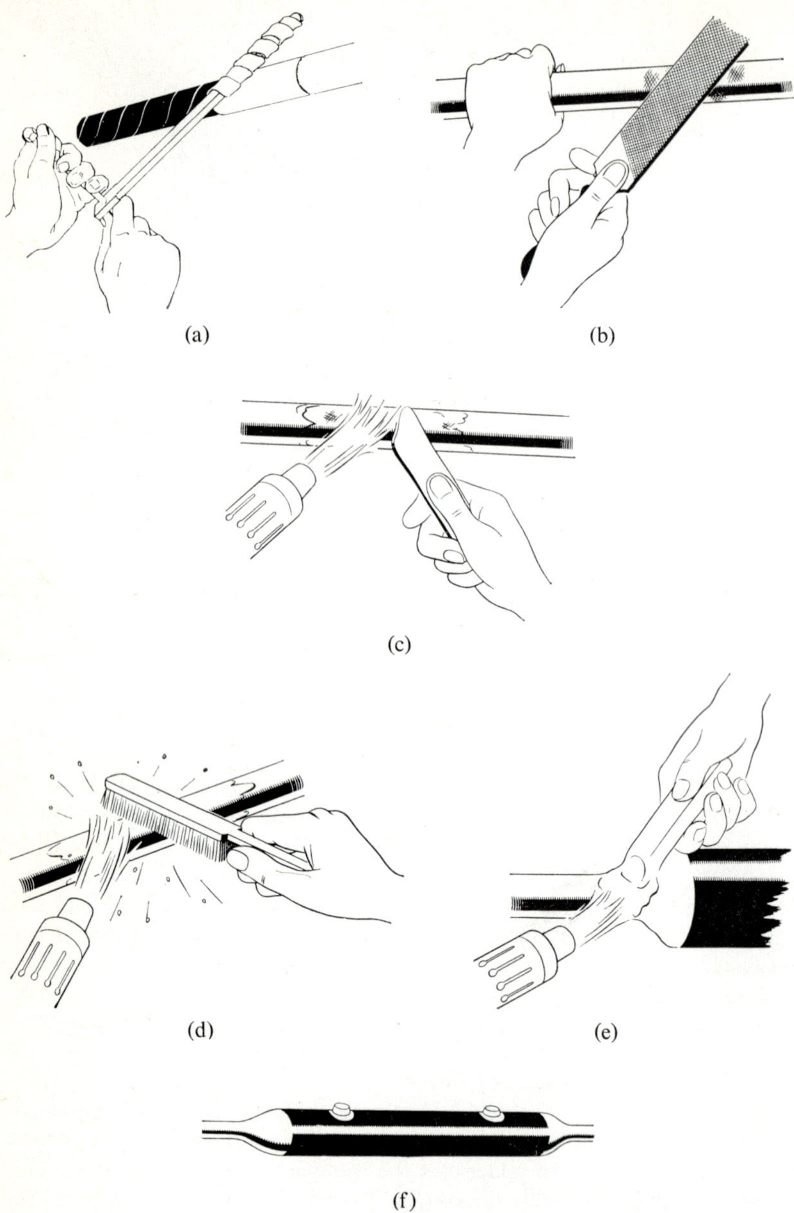

FIG. 18 *Through joint in aluminium sheathed and aluminium cored cable.* (*Courtesy Johnson & Phillips Ltd.*)

To joint the cores, the core papers are trimmed back to approximately 6 mm back from the ferrule and the remaining papers are protected by tape. Tinning is achieved by basting the core with the correct solder at a temperature of approximately 250°C. Basting at this temperature enables the core to reach a temperature sufficient for the flux to be applied with a stiff brush and is continued until the tinning process is completed. The cores are then soldered together using an already tinned ferrule. While the solder is still molten, the ferrule is tightly pinched. The joint is then insulated to the correct thickness with rubber tape.

Where the conductor is of a solid construction, compression joints would again possibly be preferred. Again, if a plastic compound is used, the application of heat is not required.

After the cores have been jointed and insulated, the box is reassembled, care being taken to ensure that the sheaths are bonded firmly to the box, after which the latter is filled with compound, and, if necessary, any bonding conductor is connected between the two cables.

At terminations, the cores may be terminated in a soldering socket, the barrel of which should be squeezed to the correct shape to accommodate the core. Alternatively, the core may be crimped into the socket by using a suitable compression tool.

If the aluminium core is to be connected directly into a copper terminal, undue mechanical pressure must not be exerted on the conductor by overtightening of any screw or clamp. If the terminal is not plated, it should be protected against corrosion by painting with a suitable material such as Densol grease.

4

Wiring systems

P.V.C.-sheathed wiring systems—conduit systems—concentric systems—prefabricated systems—grid suspension system—trunking systems.

27. *What factors affect the choice of a particular wiring system?*

Safety. Whichever type of wiring system is installed, safety must be the prime consideration. The Supply Regulations, the Factory Acts and the I.E.E. Regulations have all been formulated with this point in mind. The choice of materials and the overloading of circuits are two examples of where the safety of the installation may be impaired.

Cost. Although safety is always the first consideration, the cost of the installation often decides the type of wiring system to be installed. This exposes a weakness in the competitive system of submitting estimates, the client often accepting the lowest estimate with the result that the best wiring system is not always installed.

Flexibility. This factor is related to the economics of the system. Every consumer wants his installation to be as flexible as possible but is not always prepared to pay the additional cost. In many cases it is not even necessary, but in modern industrial and commercial installations, flexibility often takes precedence over cost.

Flexibility also includes making provision for further extensions. Therefore, when an installation is being planned, especially a large installation, provision should be made for future extensions, to ensure that the existing installation will not become overloaded.

Length of service. This must be considered in terms of long-term or short-term investment. Thus, newly erected buildings have a relatively long life and, therefore, the wiring system should also have a relatively long life. In addition to the initial cost, the annual charges must also be considered, and it is the combination of these two factors which would eventually decide which is the best system to install.

On the other hand, the short-term investor only intends to occupy the building for a short period of time, and therefore, the annual cost is secondary to the initial cost. In many instances, unsatisfactory wiring systems are installed to reduce the initial cost.

Simplicity of installation. This is related to cost in that the building structure might make it virtually impossible to install one type of wiring system and thus a different type would have to be used. Where old houses are being converted to electricity, and even in many new domestic installations, p.v.c. systems are preferred to conduit systems because of the simplicity of the installation. On the other hand, in multi-storey commercial buildings where the floors are constructed of pre-cast concrete block, conduit systems are usually necessary. For mains and sub-mains in industrial installation m.i.m.s.- or p.v.c.-armoured cables have virtually superseded conduit systems.

It is clear, therefore, that the type of wiring system often depends upon the situation in which it is to be installed.

Appearance. This is also related to cost as, in general, the more attractive the appearance, the greater is the initial cost of installation, and in many instances the annual maintenance cost. This is especially true where lighting fittings are to be installed, and, although the complete installation must conform to the occupant's requirements, the contractor must be willing to advise and discuss with him the suitability, the initial installation cost and the annual cost compatible with the finished appearance. It must not be forgotten that the more attractive the finished installation, and this includes ensuring that all fittings, accessories, etc. are installed correctly, the greater is the goodwill obtained by the contractor.

28. *Describe the precautions that must be taken when installing p.v.c. wiring systems in domestic installations.*

As the three-plate ceiling rose system would almost inevitably be installed for lighting, the method of wiring would be to take a twin and earth 1·5 mm^2 cable from the distribution board to each light in turn, then to run a twin-and-earth 1·5 mm^2 cable from every light point to the associated switch. In modern installation, the 13 A socket-outlet is almost universally employed and irrespective of whether radial or ring circuits are installed, the correct size of twin-and-earth cable must be used. Composite cables would also be used for cookers and immersion heaters, and although a diversity factor must be taken into account when calculating the size of the cables for the cooker, where the run is long, the voltage drop must also be taken into consideration. Fixed appliances may be served from spur-boxes in ring mains but it is preferable to serve immersion heaters directly from one way on the distribution board.

Precautions that must be taken when running the cables are:

Under floorboards where the cables have to cross joists, the joists should be cut to just the sufficient depth and width, and the cable should be enclosed in conduit, securely fixed, or by an equivalent means of protection. Alternatively, holes may be drilled in the joists so that the cables are at least 50 mm vertically below the floorboards.

Where the cables run parallel to the joist, they should be clipped at least every 1 m.

Where the cables are installed beneath the ground floor, normally there is no need to cut joists, but the cables should be clipped where necessary. In roof spaces, cables must not be taken diagonally from point to point, but, where they run parallel to the joists, the cables must be clipped along the side of the joist at least every 250 mm. When crossing joists, the joist may be drilled and bridges fixed between adjacent joists to which the cables may be clipped. Alternatively, the necessary length of wood may be fixed to the top of the joists, the cables then being clipped to the sides of the wood.

Where the cable is taken through the ceiling and into the accessory, the outer sheath must be removed just sufficiently for the unsheathed cores to be completely enclosed. Where the flexible cord in a lighting fitting has to be connected to the fixed wiring in the ceiling space immediately above the fitting, the fixed wiring should terminate in a joint box fixed to the joist, the connection to the lighting fitting being made by a t.r.s. flexible cord.

Great care must be exercised when removing existing floorboards and architraves; the latter should be eased off gently to avoid splitting the wood. Occasionally the floorboard has to be cut before it can be removed. A hole should be bored and the floorboard cut adjacent to the joist, after which a bridge piece should be fastened to the side of the joist so that when the floorboard is replaced it is supported solidly and level with the other floorboards.

If possible twin-cored red cables should be used from the ceiling rose to the switch position and also for the strapper wires between two-way switches. Where a cable with red and black cores are installed, red identification tapes should be provided at these positions.

In domestic installations, most of the cables are concealed, but where they are exposed, they must be clipped at distances not exceeding 250 mm for horizontal runs and 400 mm for vertical runs, although, for appearance's sake, it may be necessary to clip them at distances much less than the permissible figures. Where the cables are sunk in the plaster, they should be protected by conduit or by metal troughing. In certain instances, it may be necessary to protect exposed cables by wood casing.

Flush-mounted switches are used almost universally in domestic installations and even where they are of the all insulated pattern, an earth wire must be incorporated in case the switch is replaced by a switch with a metal plate. An earth wire must also be taken to every ceiling rose for a similar reason. Flush-mounted socket-outlets are also often preferred to the surface type, and in many situations it might be advantageous to install multi- or dual-socket outlets. Where flush-mounted socket-outlets are installed, they must be contained in incombustible boxes and the outer sheath must not be removed so far that any of the exposed cores remain outside the box. Socket-outlets should be installed at least 150 mm above floor level and in some situations, such as kitchens, it may be necessary to install them at working level. If these are surface mounted, Bakelite pattresses should also be installed.

Fixed appliances should be served from fused spur boxes, in which case, the total load must not exceed 13 A. On the other hand, they may be served by a non-fused spur served from another socket-outlet, a joint box or the origin of the circuit, in which case, the spur cables must have a current rating not less than that of the conductors forming the ring.

Where ring circuits of socket-outlets are installed, if possible, the conductors should not be cut at socket-outlets and joint boxes, but where they do have to be cut, the twisted joint must have a resistance not less than that of the unbroken cable. Care must also be exercised to ensure that the correct fuse is installed in fuseboards and plugs.

Where an immersion heater is installed, a double-pole switch must be installed to control it, and if this is required remote from the heater, possibly in the kitchen on the ground floor, a further switch must be installed adjacent to the heater, but conforming to the regulations concerning bathrooms. The p.v.c. cable must not be taken into the head of the heater, but must terminate in a joint box, from which a heat-resisting flexible cord may be taken to the heater.

Cookers should be served from an appropriate cooker control unit by cables with a rating corresponding to the rating of the sub-circuit cables, and if installed where they might come accidentally in contact with the metalwork of other services, the two services must be bonded.

29. *Describe the conduit systems available for normal installation work and state in what type of situation each would be most suitable.*

Steel metal conduit systems. These may be divided further into the following two types:

Light-gauge conduit: Light-gauge conduit comprises a light metal strip which is formed into circular or oval tubes of the required dimensions, these being normally restricted to 16 mm or 20 mm for normal installation work. The seams are either butted or brazed together, and although the conduits usually have a black enamel finish, galvanized or sheradised conduit is obtainable. Because of the thinness of the tube it cannot be screwed. To mechanically couple lengths to accessories or to each other, special grip type accessories are necessary. It is, therefore, very important when installing this type of conduit, to ensure that all measurements are accurate, so that sound earth continuity is maintained. The ends of the conduit must also be cleaned of enamel before fitting any accessory and any sharp burrs inside the conduit must be removed.

The conduit may be easily cut by filing a groove round the circumference of the conduit, then breaking it by hand. Within limits, the brazed type can be set, but it is virtually impossible to set the closed joint type. Rubber bushes are provided at outlets to prevent abrasion of the cables.

The advantages of the system are that it can be installed more quickly and economically than heavy-gauge conduit, and that a large number of cables can be installed for the same external diameter of conduit. The advantages are more than off-set by the disadvantage of the relatively poor mechanical protection afforded by the conduit, and particularly by the difficulty in *maintaining* a sound earth continuity throughout the entire life of the system. The system, therefore, has little to recommend it for lighting and power installations, but it may be advantageously installed for extra-low voltage systems.

Heavy-gauge conduit. The thickness of heavy-gauge conduit is far greater than that of the light-gauge conduit, so enabling it to be screwed, thus producing a far more efficient earth-continuity system.

The heavy-gauge welded conduit is formed by welding the seams of the tube either electrically or by acetylene, and finished with a stove-enamelled (black) surface. It is used extensively for surface work in commercial and industrial installations and to a limited degree for concealed systems also.

The solid-drawn type is heavier and superior to the welded type, being constructed without any joints throughout its entire length. It is suitable, therefore, for applications beyond that of the welded type, such as where it is to be installed in flammable or explosive situations.

For exterior and damp situations, both conduits are obtainable with either a galvanized or sheradized finish.

Aluminium conduit. This type of conduit is sometimes preferred to steel conduit, especially where there is a risk of corrosion to steel in damp situations. Other advantages are that the aluminium is lighter than steel, has a better conductivity, is relatively easy to handle, and is non-magnetic. The disadvantage of aluminium conduits is that although it is resistant to rust, it is subject to corrosion from other forms of chemical action.

Aluminium conduits are constructed similarly to steel conduits, and in one form can be handled and screwed by the same type of equipment used with the steel conduit system. A more recent type, however, has been developed which employs a sheradized grip-type fitting suitable for connecting to normal conduit accessories. This fitting gives such a strong mechanical joint that it is almost impossible to uncouple it, thereby giving sound earth continuity.

Aluminium conduit has not the same mechanical strength as steel conduit, but is sufficiently strong for most installations, provided that mechanical protection is given where necessary.

Black enamelled iron or steel fittings may be used with aluminium conduit in dry situations, but where it is installed in damp situations galvanized or sheradized fittings should be installed.

Copper conduit. Copper conduits may be of the light- or heavy-gauge type, the light gauge type being attached to accessories by soldered, capillary type fittings, and the heavy-gauge being screwed similarly to the steel conduit. The light-gauge copper conduit has the advantage in that, as the fittings are soldered to the conduit, an efficient earth-continuity system is obtained. Also, unlike the light-gauge steel conduit, the light-gauge copper conduit can be set and bent, but bending springs should be inserted to prevent kinking of the conduit. The method of installing the heavy-gauge conduit is similar to that for the heavy-gauge steel conduit system.

Non-metallic conduit. This type of conduit is obtainable in both light-gauge and heavy-gauge, circular and oval construction, a very useful form being the flexible conduit which is very suitable for concealed systems.

To ensure a continuous system, light gauge conduits are fastened together with slip-on fittings which may be sealed with a vinyl cement to give a satisfactory adhesive bond. Although heavy-gauge conduit can also be obtained suitable for grip fittings, it can also be screwed, so enabling it to be used with standard type accessories and fittings.

This type of conduits system finds it most useful application in installations where, because of corrosion problems, steel and aluminium conduits are unacceptable, such as in chemical works, platings shops, dye works, agricultural buildings, etc.

Its main disadvantages are that it has not the same mechanical strength as metal conduit systems, and its limitation of use where temperatures are liable to exceed 60° C and to fall below minus 5° C.

30. *Describe the precautions that must be taken before and after erecting steel conduit systems in order to provide good mechanical protection and sound earth continuity, and to prevent abrasion of cables.*

Mechanical protection. To ensure good mechanical protection to the cables, the conduit must be continuous throughout its entire length and for as long as it is in service. It is necessary, therefore, that the right type of conduit is selected for the conditions in which it is to be installed so that corrosion can be minimised. In industrial installations, it is sometimes necessary to protect the conduit itself where it might be damaged by heavy objects, such as trucks, striking it.

Sound earth continuity. One of the most important purposes of the conduit is to provide an efficient earth-continuity path for any earth currents that might flow in the system. A high resistance in any part of the earth-continuity conductor is a source of danger to life and property. Hence, to maintain its efficiency as an earth-continuity conductor, the following precautions must be taken when installing steel-conduit systems.

The number of conduit accessories must be limited to the amount necessary for the ready renewal and withdrawal of cables. Every joint in the conduit is a possible source of high resistance, especially where threads are liable to corrosion, or where the screw of grip-type fittings has not been tightened sufficiently, or the enamel cleaned from the conduit where it enters the fitting.

All conduits must be measured carefully when cutting for length so that unnecessary slackness is avoided where accessories are attached to conduits.

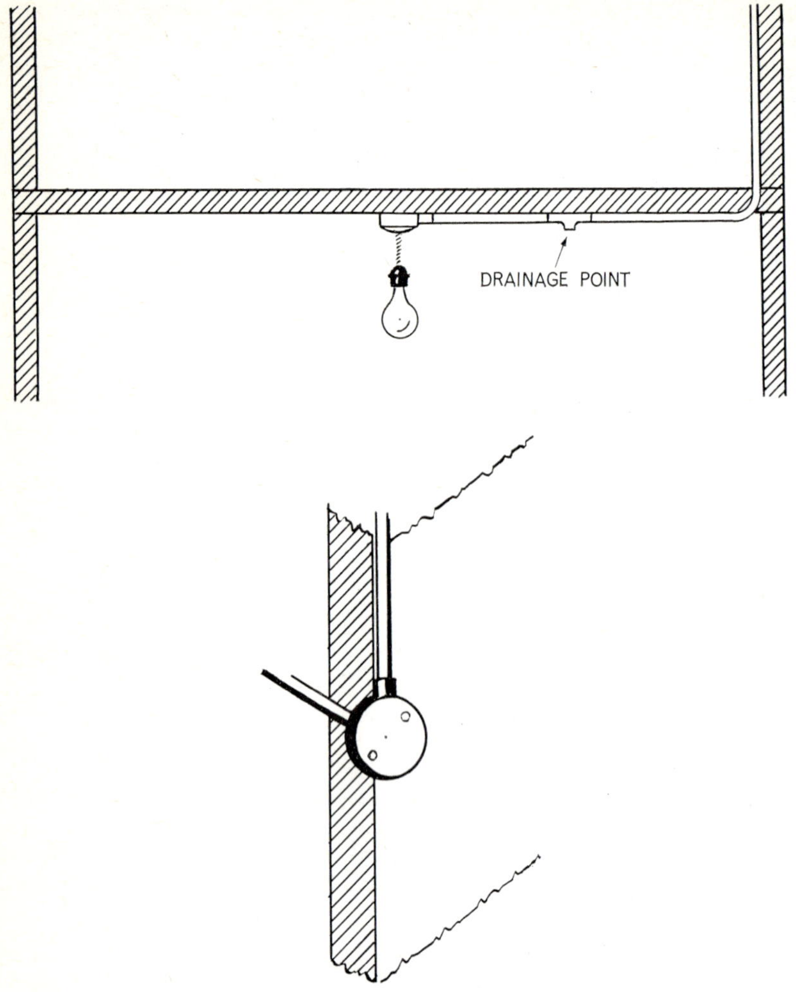

FIG. 19 *Preventing condensation in a conduit system.*

Where conduits terminate at adaptable boxes, etc., internal bushes should be screwed into couplings and, where necessary, lead washers should also be provided.

Running couplings should have locknuts fitted behind to prevent the coupling from becoming slack through vibration. Stand-off saddles should be used in exterior or damp situations so that the rear of the conduit can be kept clean from moist, corrosive dusts, and be repainted when necessary. Also, in damp situations, all threads should be coated with a good bituminous paint, as should the conduit where any enamel or galvanising has been removed.

Good mechanical support is necessary where conduits terminate at distribution boards or boxes, so as to prevent the connection becoming loose should vibration occur. Where the resistance of the earth-continuity conductor is liable to be high, outgoing conduits should be bonded to the incoming conduits at distribution boards, etc.

Condensation is also liable to reduce the efficiency of the earth-continuity system through corrosion, and where this is likely to occur, it may be minimised by installing a tee at the lowest end of the run with the spout pointing downwards (Fig. 19(a)). Condensation, however, is caused where conduits pass through varying zones of temperature, therefore to prevent condensation occurring, a back outlet box should be fitted at the junction of the two zones. (Fig. 19(b)) This is sealed with compound to prevent cold air from entering the conduit where it passes through the hot zone. A small hole is drilled at the bottom of the box to allow any small accumulation of moisture to escape.

Abrasion of cables. To prevent abrasion of cables, the following precautions should be observed.

Conduit systems must be erected complete before any cables are drawn in. Any inspection or draw-in boxes must remain accessible throughout the life of the installation so that cables may be withdrawn or additional cables may be installed. A disadvantage of some conduit installations is that inspection boxes which were originally accessible become inaccessible due to structural alterations. Accessibility also means that the boxes themselves must not only be accessible, but that sufficient room must be provided for drawing-in or withdrawing cables.

The maximum number of cables run in one conduit must be such as to permit easy drawing-in. In no circumstances must the number of cables exceed the number given in the appropriate tables of the I.E.E. Regulations (B5m; B6m). Where the size of cables or the groupings are not covered by the tables, the number of cables installed shall be such that a space factor of 40 per cent is not exceeded. These values are based on there being no more than two 90° bends or their equivalent in the run, and for short straight runs, such as switch drops, this value of space factor may be moderately exceeded. It should be noted that the space factor may be defined as the ratio:

$$\frac{\text{sum of the effective overall c.s.a. of cables forming a bunch}}{\text{the internal c.s.a. of the conduit, pipe, duct or trunking}}$$

Exceeding this space factor may not only produce abrasion of cables, but may also produce excessive temperature rises. A study of the I.E.E. Tables, B5n and B6n show that they do not cater for:

(1) Cables of different dimensions in the same size of conduit.

(2) Conduits in excess of 50 mm.

(3) Cables of size less than 2·5 mm² which are not considered suitable for conduits above 40 mm.

This is not to say that smaller cables cannot be installed in conduits larger than 40 mm, but in general it is not good practice to install a large number of small diameter cables in large diameter conduits, because of the possibility of their becoming entangled when being drawn in.

The radius of any conduit bend must be such as to fulfil the requirements of the appropriate regulations concerning the bending of cables, and in any event, the inner radius of the bend must not exceed 2½ times the outside diameter of the conduit. It should be noted, however, that the radius of the conduit bend is dependent not only upon the diameter of the cable, but also upon the type of protective sheath used.

Solid elbows or tees must not be used except immediately behind a lighting fitting, outlet box or accessory. A solid elbow may be installed at a position not more than 500 mm from a readily accessible outlet box in a run not exceeding 10 m provided that all the bends are not equivalent to more than one right angle bend. Where solid elbows or solid tees are installed, it may be necessary to reduce the number of cables in the conduit to facilitate the renewal or withdrawal of cables.

Ends of lengths of conduits must be reamed and where they terminate at boxes, etc., those not fitted with spout entries must be bushed. Note that for minimising the abrasion of cables, either external or internal bushes may be used. In some instances where the conduit is left unscrewed, rubber bushes may be fitted.

Boxes with ample capacity must be installed at every junction involving a cable connection and, where necessary, at terminations. At terminations, the boxes must be constructed of non-flammable non-absorbent material.

Although the loop-in system of wiring is usually employed, it is often possible to save a considerable amount of cable by installing joint boxes at appropriate places in the system. The box must be of ample capacity to contain the finished joint, which should be such that it can be fastened rigidly in the box. Because of the necessity of maintaining sound earth continuity in conduit systems, metal boxes are normally preferred to hardwood or plastic boxes.

Where cables are being drawn into conduit, they should be run off the reels or drums, care being taken that no damage is suffered by the cables while they are in movement across the floor. Alternatively, the reels may be arranged to rotate freely on a roller supported on a stand, the reels being so positioned that the cables do not become trapped between adjacent reels. One method that must not be adopted is to draw the cables from stationary reels.

No cables should be drawn into conduits likely to contain moisture. In damp situations, the conduit, if necessary, should be dried by drawing a cotton swab attached to a draw wire through it.

In short runs and where deep sets are avoided, it is usually possible to thread a draw wire through the conduit, and, in certain instances, the cables themselves. In long runs, or where the conduit contains deep sets, a steel tape should be threaded through the conduit to which the draw-wire is attached.

It must also be ensured that the draw-wire is sufficiently large and that the cables are securely attached to it. Where several cables are being drawn in, they must be fed straight into the conduit to avoid twisting the cables inside the conduit and also to ensure that they are not dragged against the side of the box when being drawn in, using a cable comb where necessary. Where the ambient temperature is high, so tending to soften the insulation, rubbing french chalk on the cables will assist the drawing-in process.

Fig. 20 illustrates several types of conduit fitting.

31. *Describe the advantages and disadvantages of an insulated conduit system, and state where such a system might be installed.*

The advantages of the insulated conduit system are:

(1) It is generally unaffected by water, acid and many other corrosive elements.

(2) It is weatherproof, can be exposed to sun, sea and general atmospheric conditions and can be buried in concrete or plaster without deterioration from corrosion.

(3) Where earthing of appliances is required, a separate earth-continuity conductor must be drawn into the conduit. This is an advantage in that the earth continuity conductor is protected against mechanical damage and its resistance is not affected by the corrosive elements which affect metallic conduit systems.

(4) It is non-ageing, light to handle and will not support combustion.

(5) Smaller sizes can be telescoped into the next larger size.

(6) No condensation problems are encountered and therefore the capacity of the conduit is higher than that of the corresponding size in steel conduit.

The disadvantages are that, although the conduit has quite a high resistance to mechanical pressure and shock, it is not so strong as steel conduit. If, however, mechanical protection is provided where necessary, it can be installed in conditions which were once considered unfavourable for this type of conduit.

Another disadvantage is that the conduit tends to become brittle at very low temperatures and to soften at relatively high temperatures. It should not therefore be installed in temperatures less than 15° C or in excess of 60° C.

Light-gauge conduits are erected in a continuous length by fastening lengths to one another and to accessories by slip-on fittings and sealing them with a vinyl cement. The heavy-gauge conduit may also be erected using the same type of fitting, but it can also be screwed so that it can be used with standard accessories. Screwing non-metallic conduits is much simpler than screwing steel conduit, but new and sharp dies are essential, and if necessary, Vaseline may be used as a lubricant. The conduit can also be bent quite simply, by pre-heating to a temperature of approximately 110° C for about 150 mm each side of the bend. A bending spring should be inserted to assist the bending operation.

Non-metallic boxes must not be used for suspending a lighting fitting where the temperature is liable to exceed 60° C or where the weight exceeds 3 kg. The conduits must be well supported to allow for the expansion and contraction where it is installed in situations of variable ambient temperature.

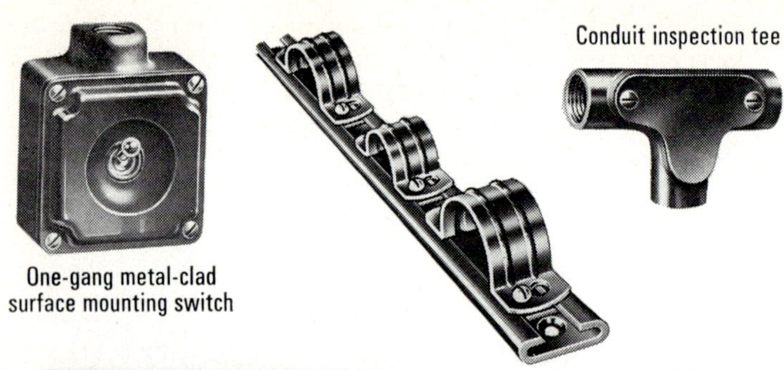

Conduit inspection tee

One-gang metal-clad surface mounting switch

Spacer bar conduit saddle

Conduit sleeve

Part sectionalised view of a through conduit box

Angle conduit box

Types of bushes

Conduit locknuts

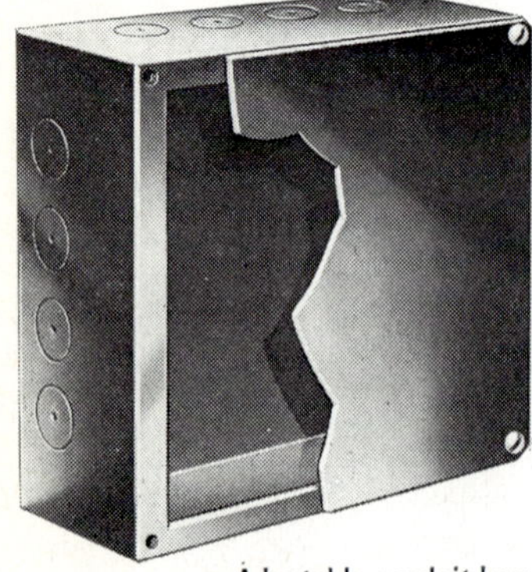

Conduit inspection elbows

Adaptable conduit box

FIG. 20 *Types of conduit fittings. (Courtesy Walsall Conduits Ltd.)*

32. *Describe the hazards associated with an earthed concentric wiring system and state what precautions must be observed to minimise them.*

In this type of system, now associated with mineral-insulated cables, the outer sheath is used not only as the earth-continuity conductor but also as the neutral conductor. This means that a conductor which can be touched is carrying current and therefore, if a break should occur in it, a dangerous potential could possibly occur across the break, or any high resistance in the sheath. To minimise the possibility of anyone receiving a severe shock, therefore, stringent precautions are necessary.

One of the Supply Regulations stipulates that the standard a.c. system must be earthed at one point only, and, therefore, before such an earthed concentric system can be installed from a public supply, authorisation must be obtained from the Ministry of Power. Otherwise the system must be served from a transformer or convertor which has no metallic connection with a public supply, or from a private generating plant.

To minimise the possibility of a dangerous potential across a break or high resistance in the external conductor, the conductor must be frequently earthed so that, should such a break or resistance occur, a parallel path is provided through which the current can flow. As this cannot be guaranteed to be completely effective, a further precaution is to incorporate a bonding conductor across every joint in addition to the means of sealing and clamping the external conductor. Even then, the system is not 100 per cent safe, and it seems doubtful if any system can be perfectly safe where the current-carrying conductor can be touched throughout its length.

For this same reason, it is important that no fuse, non-linked switch or circuit-breaker is inserted in the earthed external conductor.

As one core of any flexible cord has to be connected to the external conductor, it is essential that a sound terminal connection is provided in every box or accessory where such a connection has to be made.

33. *Describe the wiring systems suitable for a newly erected block of flats.*

If a withdrawable system is required, then either a steel conduit or plastic conduit system may be installed using p.v.c.- or rubber-insulated-cables. A plastic conduit system is usually economical and satisfactory to install, especially where the flexible type is used, but its applications are restricted, and in damp or hot, humid situations only galvanized heavy gauge steel conduit is permissible. Where underfloor heating systems are installed even this system is not very satisfactory.

For a non-withdrawable system, mineral-insulated copper-sheathed cables may be installed either with or without a p.v.c. covering. This system is particularly useful where high ambient temperatures are liable to occur in the vicinity of the cables, such as where a floor heating system is installed. A further advantage is that, where a large number of similar flats are to be erected, the wiring system for

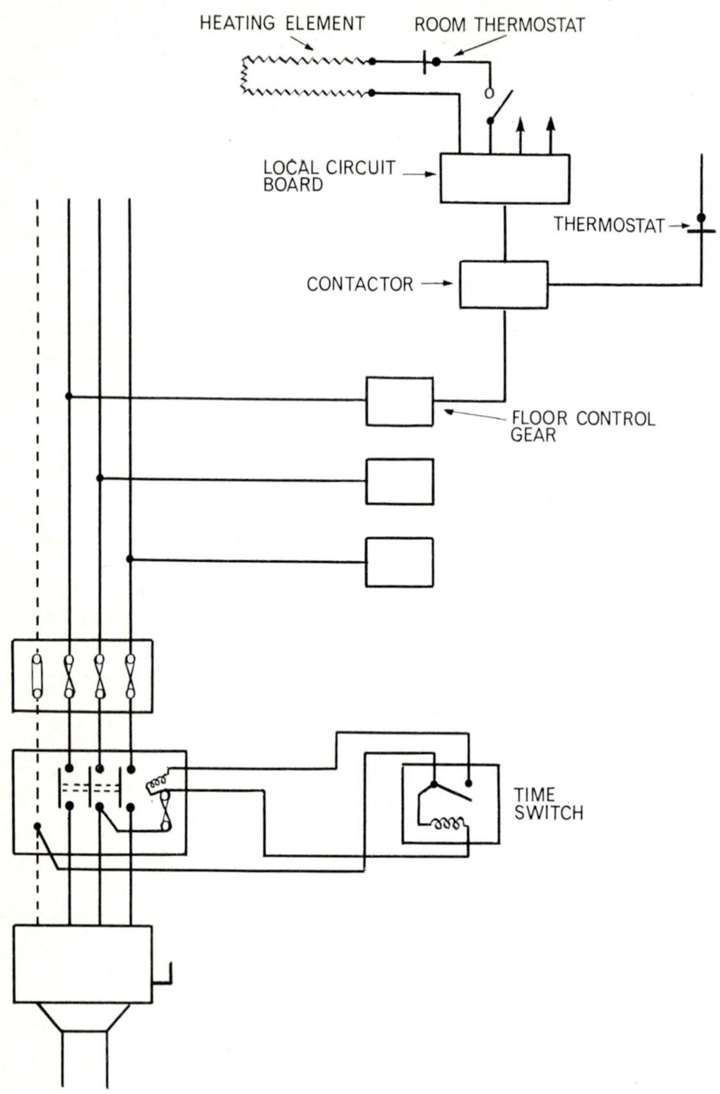

FIG. 21 *Underflooring heating system for blocks of flats.*

each flat may be fabricated in the workshop so simplifying the installation on site. In fabricating such a system, it is important to ensure that any individual length is not cut too short, but any slight excess in length may be concealed in the wall plaster.

In certain areas and provided that the necessary authorization from the Ministry of Power has been granted, the m.i.c.c. system may be in the form of an earthed concentric system.

34. *Describe how an underfloor heating system may be installed in a block of flats where the heating is part of the rental service.*

Because of the nature of their floor construction, modern multi-storey blocks of flats lend themselves admirably to the underfloor system of heating, thereby taking advantage of the off-peak tariff. Where, however, the heating is part of the rental service, there is no necessity to install local time-switches and meters. Instead a rising main may be taken vertically up the building, controlled at the lower end by a time-switch-operated contactor and possibly a thermo-time regulator, the thermostatic part operating on the external variations of temperature.

As the individual floors would be served from a single-phase supply, it is necessary to balance the floors on the rising main, the system being illustrated in Fig. 21. Although Fig. 21 shows a system suitable for large blocks of flats, for smaller blocks some modifications would be necessary, such as in the types of mains to be installed.

Normal practice is to install sufficient cables to enable a loading of approximately 140 W/m^2 to be obtained. Where several rooms have to be heated the current limit is about 13 A per circuit. Several sub-circuits may be necessary, so requiring the installation of a distribution board. Local thermostats should also be installed in each room, and at least one switch must be installed to control the heating system in each flat.

35. *Describe the advantages of the 'Octupus' wiring system and state how it is installed.*

This is a prefabricated system and applies specifically to estates of similarly constructed houses. Most of the work is done off site in the factory, so reducing the number of man-hours on site, and thereby reducing the installation costs considerably. In this system, a central joint box is provided on each floor. Flexible cables with a strong metallic braid are attached to the individual accessories, these being cut to the required length and then labelled and delivered on site in kit form with all the necessary fixings and instructions. The cables are then secured to the joint box, and, as such, a complete harness is formed embracing all the necessary connections for the building.

Fig. 22 illustrates the 'Octopus' system of wiring.

FIG 22 *Octupus system of wiring.*

36. *Describe the duct tube system of wiring and state how it is installed.*

This system (Fig. 23) consists of containing the cables in ducts formed by inflating rubber tubing with a braided fabric core which prevents adhesion to the concrete when the tube is deflated. To form the ducts, the tubing is laid along the required runs and inflated before the mixed concrete is applied. Twelve hours after the concrete has been applied the tube is deflated and removed, leaving a virtually smooth concrete duct. Provided that reasonable care is taken in handling the tubing, it can be used for several further applications. The tubing may be kept in position before and after pouring by passing it through pre-cast concrete blocks of cement fixed to the shuttering or reinforcement.

When inflated, the tube diameters range from 20 mm to 125 mm, the smaller sizes being suitable for the cables of wiring systems, and the larger sizes for main cables. Tubing over 20 mm may be inflated slightly in excess of its nominal diameter, the pressure varying between 50 g/mm^2 and 56 g/mm^2. The tube is normally inflated by means of a small compressor or air cylinder, although small lengths may be inflated by a motor-car foot pump. The tube loses approximately 10 per cent of its length when inflated which must be allowed for when calculating the required length.

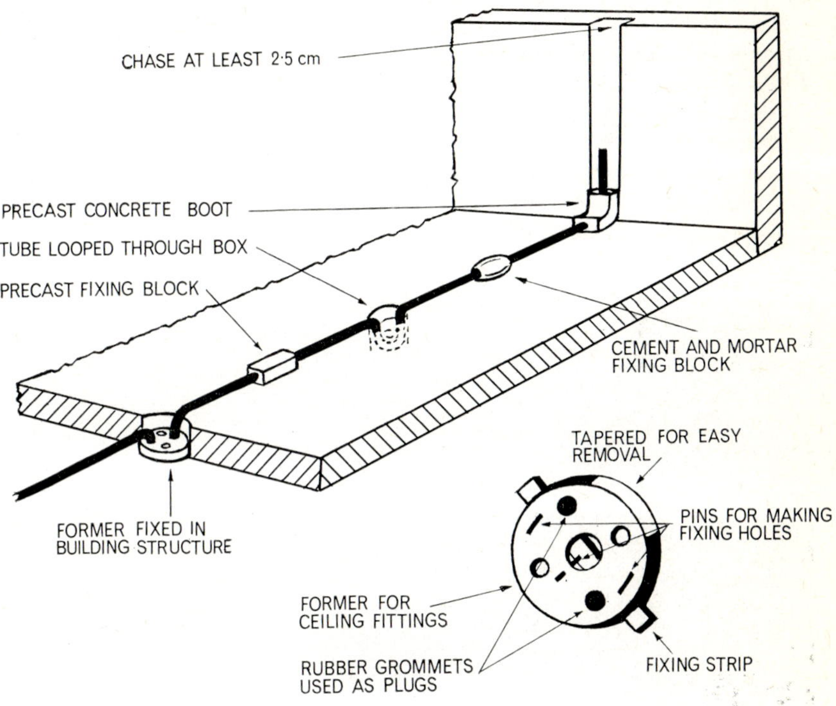

FIG. 23 *Duct tube system of wiring.*

Wooden formers may be used instead of conduit boxes at outlets, the tubing being taken through the necessary holes in the formers, the conduit boxes being attached to a notched steel strip in the shuttering. When the concrete has set, the former may be withdrawn leaving the strip with the necessary fixing holes and a suitable cavity for the cables.

To avoid wrinkling of the tube, the radius of any curve in the tubing should be not less than four times the inflated diameter.

37. *Describe the 'Grid' suspension lighting system and state how it is installed.*

Where difficulty is experienced in serving and suspending lighting fittings in industrial installations with apex roofs, a catenary wire is often used, which is stretched from wall to wall, to which the wiring system is attached and from which the lighting fittings are suspended. The grid-suspension system dispenses with the separate catenary wire and comprises two, three, four or six insulated cores, ranging from 1·5 mm² to 70 mm² laid helically round a 7/1·70 mm high-tensile galvanised-steel catenary wire. The conductors are bedded with fillings of jute or hemp to produce a circular construction, and are then encased with a braid of jute saturated with a lead-base compound which, when waxed, gives a waterproof finish. If required, additional protection may be provided by enclosure in a fire-resisting plastic or metal braid.

Special types of box are required at light outlets, to which the cable conductors are connected in the form of terminal, right angle, through, tee or four-way connections. The box may also be used to anchor the catenary wire and contains an earth terminal to which the catenary wire is fastened. It should be

FIG. 24 *Grid suspension lighting system.*

noted, however, that should the resistance of the catenary wire be too high, a separate earth-continuity conductor may also have to be used. A central bolt allows the anchoring clamp and the light alloy covers to be attached, and also enables brackets, chains or straining fittings to be coupled to the box.

In addition to having a straining fitting, where the run is long, it is advisable to install auxiliary suspension points at intervals between the main anchorages to prevent sag in the cable.

The connection boxes are designed so as to eliminate the necessity of cutting the conductors at connection points, the box being divided so that it can be fitted over the conductor, bushes also being provided to prevent abrasion of the cables. Standard line-tap connectors are then used to make the necessary connections to outgoing cables. A specific advantage of this system is that it can be prepared at floor level before being erected. Before erection, however, it is essential to ensure that the main straining anchorages are correctly aligned, and that all fixings are capable of withstanding the strain imposed by the taut catenary wire. Fig. 24 illustrates the system installed in an industrial installation.

38. *Describe the advantages of a metal trunking system in an industrial installation and state what precautions must be taken when installing such a system. How may the number of cables to be installed be calculated?*

Trunking systems are often installed in addition or in preference to conduit systems where a certain degree of flexibility is required, such as where a large number of small diameter cables have to be installed, which does not warrant the installation of more expensive wiring systems. They are also used where cables too large to be drawn into conduit have to be installed. Fig. 25 illustrates how conveniently a length of trunking may be installed to serve a row of machines in place of individual conduit runs.

In the type illustrated in Fig. 25, the trunking is constructed in 3 m lengths of zinc-coated sheet-steel, special bridge pieces being used for connecting the lengths together which also provide fixings for the cover and act as cable retainers.

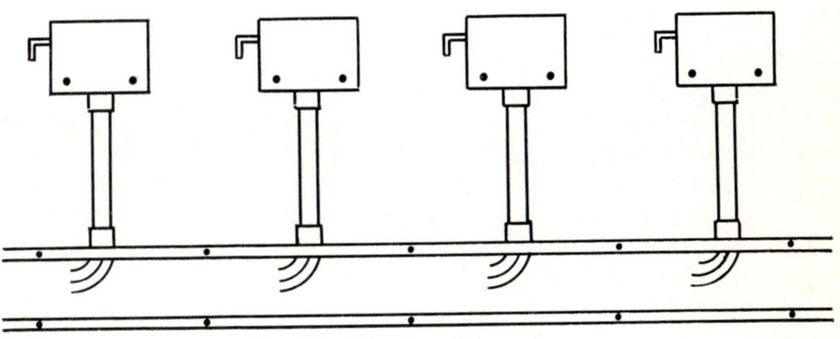

FIG. 25 *Metal trunking system.*

When installing such a system, therefore, the following precautions must be taken.

As the trunking is used as the earth-continuity conductor, it is essential that the earth continuity is effectual. At connection points, therefore, areas adjacent to the fixing screws are left unpainted; if necessary copper bondings links may be provided.

Where necessary, trunking must be protected against mechanical damage and must be securely fixed.

In long vertical runs of trunking, pin racks should be fitted at intervals of not less than 5·0 m to support the cables and to prevent undue compression of the insulation at the top of such runs. Also, in long vertical runs ventilation barriers must be fitted at intervals of not less than 3 m. Where the trunking passes through walls, floors, etc., suitable internal fire barriers must be provided to prevent the spread of fire. Also where the trunking passes through walls, floors, etc., the hole must be made good with cement or other fire-resisting material.

Where a trunking has to contain cables operating at low- and extra-low-voltage, where the categories of insulation are different, the cables must be effectively segregated by a dividing fillet securely fixed. This may entail the fitting of complicated cross-over boxes at interconnection points to ensure segregation of categories, especially where four-way boxes have to be fitted.

Trunking must be prevented from coming into contact with the metalwork of any wiring system operating at extra-low-voltage, or any radio, telephone, bell or call circuit or other system outside the Regulations, or the metal pipes of other services, for example, gas or water. If this is impracticable, the metalwork of both systems must be bonded. This is so that accidental contact between the two services cannot produce a spark which might puncture the sheath of a gas or water pipe.

The trunking system must be routed so as to prevent the entry of water into the finished trunking, and every outlet should be bushed. No cables should be drawn in, until the trunking system is completely erected.

Where cables of a.c. systems are installed, they must be bunched so that the cables of all phases and any neutral conductor are contained in the same trunking. This Regulation is sometimes broken where two conduits leave the trunking side by side, the outgoing cables being contained in one conduit and the ingoing cables in the other. The induced currents in the metalwork produces an excessive temperature rise in the cables.

The number of cables in the trunking must be such that a space factor of 45 per cent is not exceeded.

Where a large number of cables of various sizes have to be contained, it sometimes becomes laborious to calculate the size of trunking required, but if the unit system is adopted, the calculation is considerably simplified. The method consists of applying appropriate rating factors (given in Table 1) for the 600/1000 V grade cable, either p.v.c. or rubber-insulated, then relating the total number of units to the permissible number of units for the size of trunking to be installed.

Let us assume that 40 2·5 mm^2 600/1000 V p.v.c.-insulated cables, and 40 4·0 mm^2 600/1000 V p.v.c.-insulated cables are to be installed in trunking.

Referring to Table 1, the multiplying factor for the 2·5 mm² cables is 23 and for the 4·0 mm² cables is 33. Hence the total number of units is $(40 \times 23) + (40 \times 33) = 2240$. Relating the number of units to trunking sizes, the nearest size is found to be either a 100 mm by 40 mm or alternatively a 75 mm by 50 mm trunking. Actually this trunking size is capable of accommodating 2700 units.

Table 1. *Unit system for calculating the dimensions of trunking*

Cable size	*Factor*		*Trunking size (mm)*	*Capacity/units*	
	V.R.I.	*P.V.C.*		*Unlaced*	*Laced*
mm²	600/1000 V	600/1000 V			
1·0	25	15	40 × 40	1000	830
1·5	36	23	50 × 50	1800	1480
2·5	39	26	75 × 50	2700	2200
4·0	55	35	100 × 50	3600	2900
6·0	65	44	75 × 75	4050	3300
10·0	76	53	100 × 75	5400	4400
16·0	95	69	150 × 50	5400	4400
25·0	118	82	100 × 100	7200	5900
35·0	194	142	150 × 75	8100	6660
50·0	300	225	150 × 100	10800	8880
70·0	443	315	150 × 150	16200	13300
95·0	565	408	230 × 100	16200	13300

39. *Describe a rising mains system suitable for:*

(1) *A multi-storey commercial building,*

(2) *A twelve-storey block of flats.*

What precautions must be taken when installing such a system?

(1) A type of system used extensively in commercial buildings is where bare conductors are installed in earthed-metal trunking or in a chase contained in the structure of the building, the trunking or chase extending the full height of the building.

Assuming that an earthed metal trunking system is being installed on a standard a.c. four-wire supply, four bare copper or aluminium conductors of a cross-section appropriate to the current to be carried are held suitably spaced on insulators mounted inside the trunking. Standard lengths are 4 m, but intermediate lengths are sometimes necessary. The lengths are mechanically coupled similar to the overhead busbar system, copper earth-bonds being provided to ensure sound earth continuity. Expansion joints of laminated copper are also provided in the conductors in long vertical runs, and sliding supports are fitted at intervals to allow for the longitudinal movement of the bars.

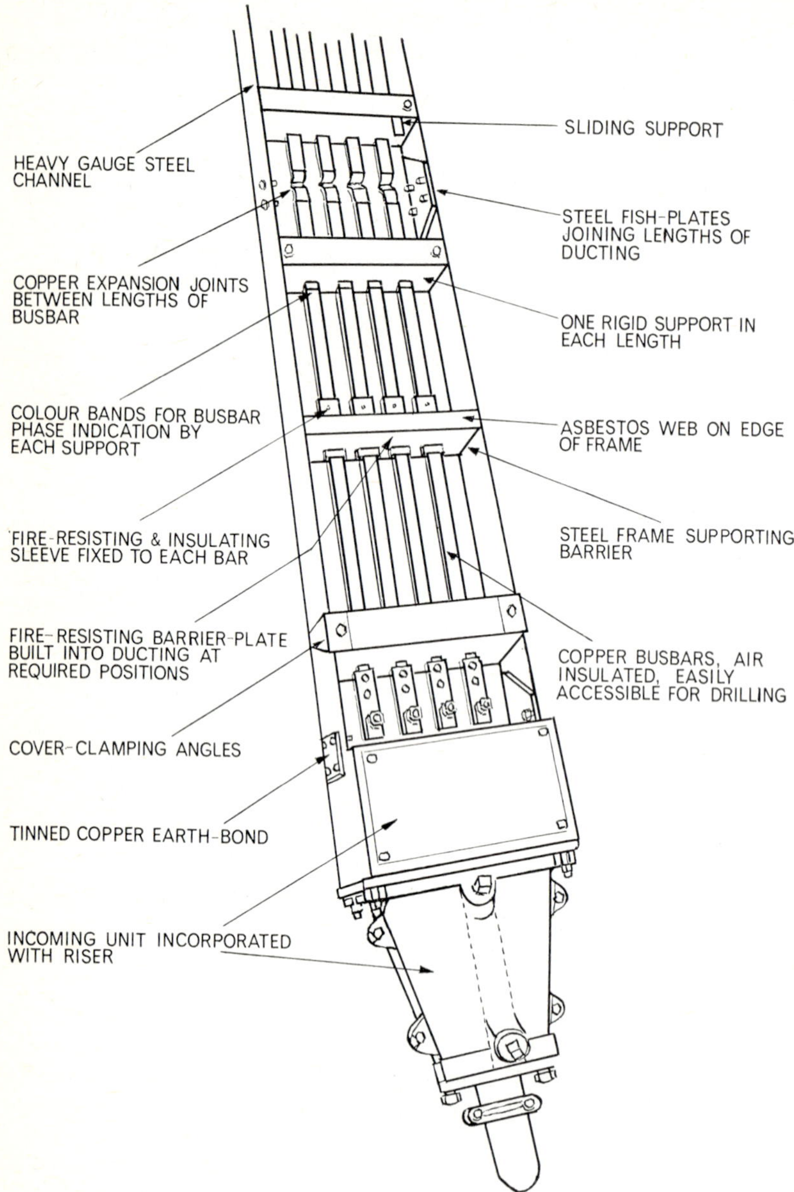

FIG. 26 *Part layout of rising mains system.*

The incoming cable unit is invariably situated at the lower end of the system, and, although Fig. 26 shows it arranged for the connection of a paper-insulated cable, provision can be made to accommodate other mains systems.

The busbars may also be of a circular or rectangular section, the latter having the higher current rating for the same cross-section.

FIG. 27 *Wall-mounted distribution system for rising mains.*

The control equipment on each floor is dependent upon whether the floor area is sub-divided into a number of tenants or whether the floor area is occupied by a single tenant. Assuming the latter, the control equipment may consist of either a floor or wall-mounted distribution unit (Fig. 27), which is an integral part of the system, and which is capable of controlling the entire section. The unit would almost certainly be of a three-phase construction and be fitted with an incoming control equipment and the necessary outgoing fuses or circuit-breakers. If single-phase units are installed, the floors should be progressively balanced over the three phases.

The busbars are protected from unauthorised entry by front detachable cover plates, any fire barriers being indicated by labels at the appropriate place. Where appearance is essential, the trunking is contained in a duct or chase, and may be fixed by external side brackets or by holes in the trunking.

Where the building is divided into small tenancies, it is necessary to make special provisions for controlling and metering the landlord's services normally at the intake position.

(2) For blocks of flats, because of the cost, other rising mains systems would be preferred to the bare conductor system in trunking. One such system employs a p.v.c. armoured cable which runs, normally in a duct, the full height of the building and controlled by a sealed cut-out unit at the lower end. Special connection boxes are installed at floor levels, which serve as a joint box in the mains and a distribution centre to the flats on the associated floor.

Each tenant has his own consumer's unit, and where the meters are mounted in the control unit a neat appearance may be obtained by installing a flush-mounted unit. In many cases, it may be more convenient for all the meters on the one floor to be grouped in the one position, so that they can be read without entering any of the flats.

Where several flats on one floor are being served from one rising main, a distribution board, which may be an integral part of the system may be installed. This board would contain the service fuses and links, and would be sealed. Lateral sub-services would then be taken from the board to the individual consumer control units. For small blocks of flats, this may not be necessary, and the consumer unit itself may be used as a terminal block in the rising main.

Although a three-phase and neutral cut-out may be used to control the rising main, Fig. 28 illustrates an alternative arrangement whereby three single-phase and neutral rising mains are served from single-phase and neutral switch-fuses or fuse-switches mounted on a sealed busbar chamber. The individual floors would be balanced over the three phases, by connecting each rising main to every third floor, respectively.

A landlord's service must also be provided and where a busbar chamber is installed, a switch-fuse and meter would be mounted on the chamber for such a service, while where a three-phase rising main is installed, a three-phase and neutral fuseboard may be installed near the intake position.

When installing such systems, the following precautions must be taken: When passing through floors, fire barriers must be provided (Fig. 26); the

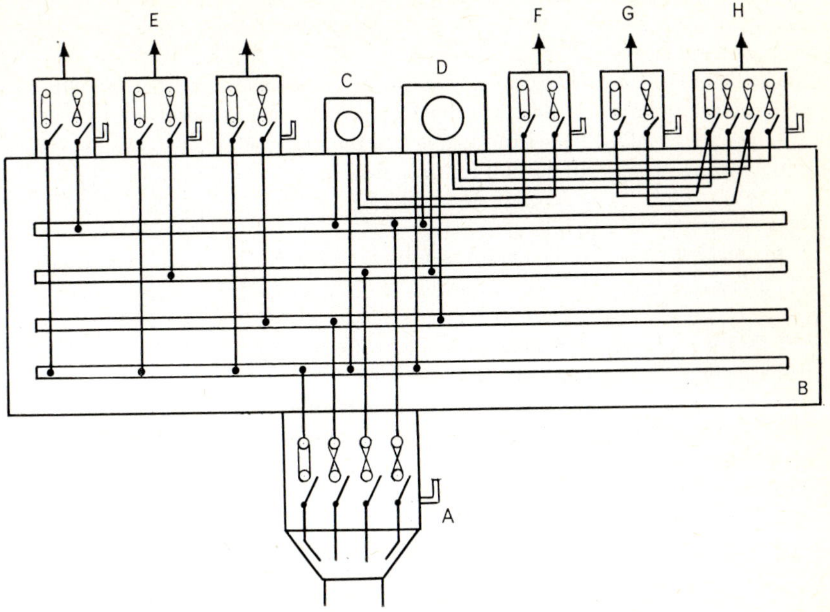

A. SEALED T.P. AND N. SWITCH FUSE
B. SEALED T.P. AND N. BUSBAR PANEL
C. LANDLORD'S SINGLE-PHASE LIGHTING METER
D. LANDLORD'S THREE-PHASE POWER METER
E. SEALED SINGLE-PHASE SWITCH FUSES
F. LANDLORD'S SINGLE-PHASE LIGHTING SWITCH FUSE
G. LANDLORD'S SINGLE-PHASE POWER SWITCH FUSE
H. LANDLORD'S THREE-PHASE POWER SWITCH FUSE

FIG. 28 *Illustrating the use of single-phase rising mains.*

conductors passing through a slotted plate with fire-resisting sleeves. As these would serve in place of ventilation barriers, there is no necessity for fitting the latter.

The conductors must be inaccessible to unauthorised persons and fixed in earthed-metal trunking or in a trunking or shaft, specially provided for the purpose. They must be of adequate strength to withstand the electro-mechanical forces that may be set up by the prospective short-circuit current.

Although the force exerted between the conductors may be relatively low when they are carrying their normal current, under short-circuit conditions the force exerted may be such as to impose excessive strain on insulators and conductors if they are placed too close to one another or are insecurely fixed in the trunking.

The I.E.E. Regulations specify a minimum distance of 75 mm between circular conductors and 75 mm or 30 mm for rectangular risers when disposed in parallel or edge-to-edge, respectively.

They must be free to expand and contract as the temperature changes without detriment to themselves, or to any other part of the installation (Fig. 26).

Where the trunking passes through floors, the surrounding hole must be made good with cement or similar fire-resisting material.

40. *Describe how a cable-tap wiring system is installed and state its advantages and disadvantages.*

This system is installed where a number of machines have to be served. The trunking may be attached to the walls or suspended overhead. Instead of a large number of small size cables being served individually to each machine, relatively large cables are contained in the trunking, the number depending upon the type of system to be installed.

In the system illustrated in Fig. 29 there are three or four 50 mm² p.v.c.-insulated cables served from a 60 A three-phase and neutral switch fuse, the trunking being 50 mm by 50 mm in cross-section. At pre-determined points

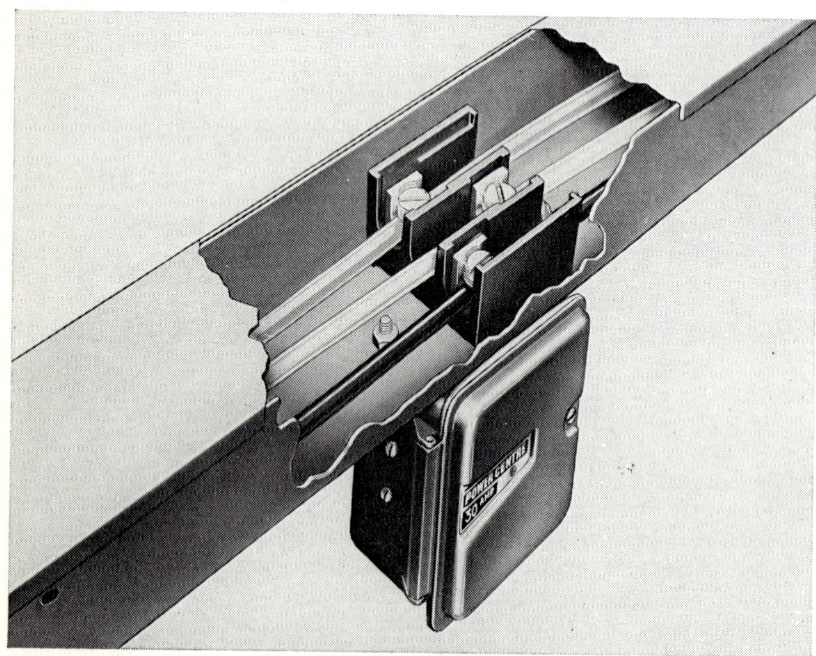

FIG. 29 *Cable-tap wiring system.*

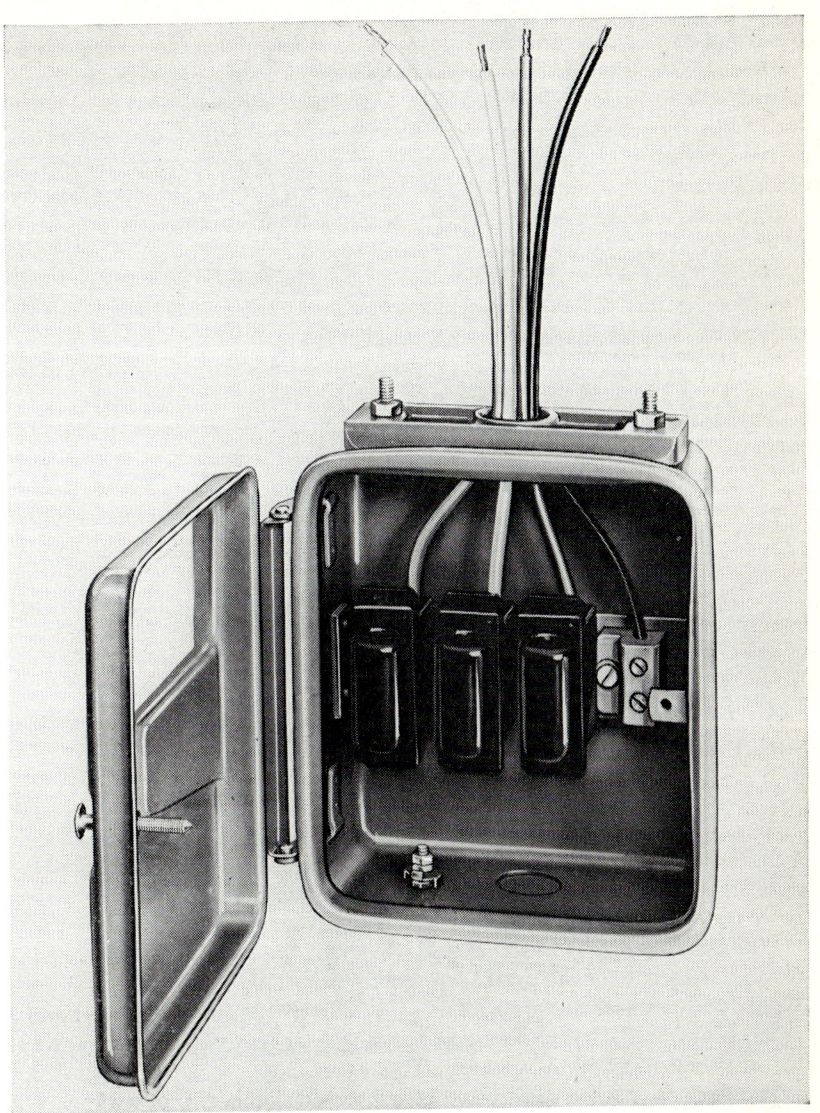

FIG. 29(b)

connector blocks are incorporated in the circuit to which the main cables are connected. Connections are then taken to tapping-in boxes mounted beneath the trunking and adjacent to the machine to be served. The boxes contain suitable fuses and possibly a neutral link, and also knock-outs so that conduits, etc., may be taken to the machine being served. Trunking of a plain construction may be installed, but to give a higher degree of flexibility, the trunking may be obtained with holes every 650 mm apart, the holes being covered with blanking-off plates.

Alternatively a 75 mm by 75 mm trunking may be installed containing possibly four 95 mm^2 p.v.c.-insulated cables served from a 100 A three-phase and neutral switch-fuse. If the system is arranged to be served from the centre, the capacity of both systems may be doubled.

The trunking is normally obtained in 3 m lengths with provision being made for coupling adjacent lengths together by means of a sleeved coupler, while the cover may be arranged for either top or front fixing.

It is also possible to accommodate cables for other services in addition to the main cables, provided that the space factor for trunking is not exceeded.

The advantages of such a system are that it gives a higher degree of flexibility than the normal trunking system and only requires the one means of control.

On the other hand, its disadvantages are that, should one of the main fuses blow, the entire system has to be disconnected while a replacement fuse is fitted. Also, since the system has to be disconnected while additional tapping-in boxes are fitted, it is not so flexible as the overhead busbar system, but alternatively has the advantage of being the more economical system to install, especially where a number of small powered machines have to be installed.

41. *Describe the advantages of the overhead busbar system and state how it is constructed and installed.*

The overhead busbar system is used almost exclusively in industrial installations, particularly where the load density inside the factory is at its highest value. Thus, its main application is where lines of machines require a service, but where a possible interchanging of positions requires a flexibility not afforded by any other system.

Where the machines are arranged close to a wall, the overhead busbar trunking may be fastened to the wall and conduits taken from it to the individual machines. Where the machines are arranged in lines remote from the walls, the normal practice is to run the busbars overhead, possibly attached to the cross-members of the structural steelwork by suitable brackets (Fig. 30).

The busbars may be constructed of copper or aluminium, of a circular or rectangular cross-section, and are mounted on insulators in metal trunking normally in 3 m or 4 m lengths. The busbars and trunking are bolted together to form a continuous length over the entire length of the machines to be served. A connection box is mounted, possibly at the side of the trunking and is designed for accommodating conduit, m.i.m.s., paper or p.v.c.-insulated cables which

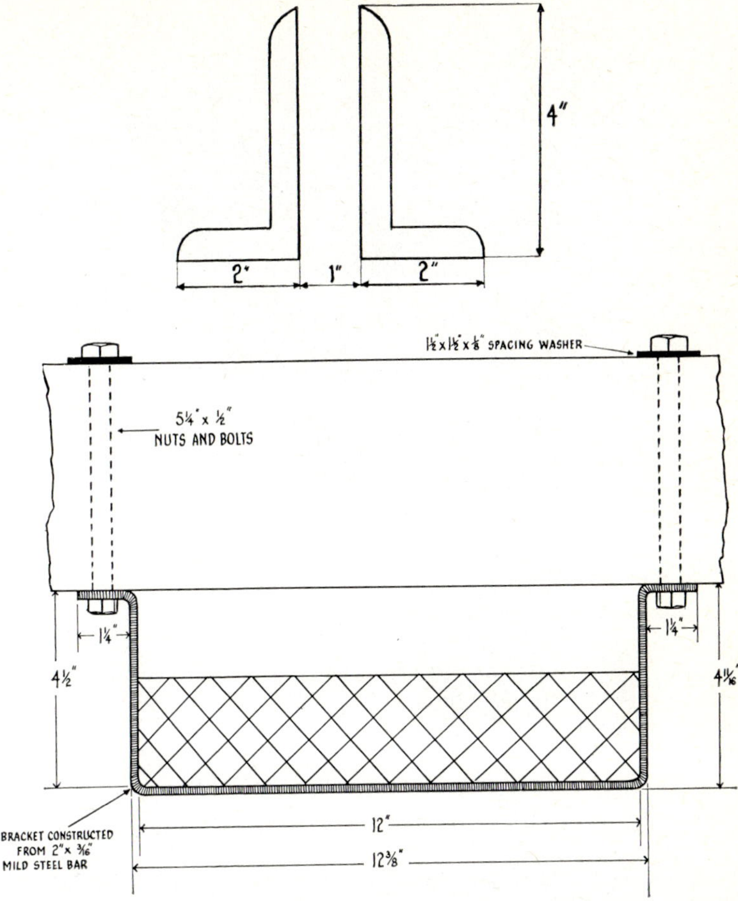

FIG. 30 *Overhead busbar system attached by brackets.*

serve to connect the busbars to the supply. The box may be mounted at either end, but should preferably be mounted at the centre. Fig. 31 illustrates one method of connecting the supply cable to the trunking.

The busbar ratings range from the miniature size of 100 A to 1000 A and tapping positions are provided at regular intervals. It is essential that the tapping boxes can be connected to the busbars while the latter are alive. Fig. 32 shows how easily this is accomplished. The boxes incorporate the necessary fuses and link for controlling the outgoing cables, the fuses being of the h.b.c. type and normally rated at 15 A, 30 A, 60 A or 100 A respectively. Sound earth continuity is assured by screwing two retaining rods into nuts inside the trunking, the contact being assisted by means of phosphor-bronze earthing springs.

FIG. 31 *Illustrating mains units for overhead busbar systems.*

By having tapping points at such short intervals, the maximum degree of flexibility is ensured, the loads being positioned so that only a short run is required from the trunking to the machine.

Where lengths of trunking are coupled together, both lengths are mechanically coupled by nuts and bolts, a copper strip also being provided to ensure sound earth-continuity. The busbars are normally connected together by copper fish-plates, but in long runs and in normal temperatures, at intervals of approximately 50 m, the fish-plates should be replaced by expansion joints which may comprise either spring-loaded rectangular section conductors in a telescopic formation or a flexible laminated expansion joint (Fig. 26).

To obtain the most satisfactory operation from such a system, the lines of trunking should be suspended from the cross-members of the structural steelwork or alternatively should rest on the cross members, but in each case, the trunking should extend the full length and be situated as near as possible to the line of machines with which it is associated.

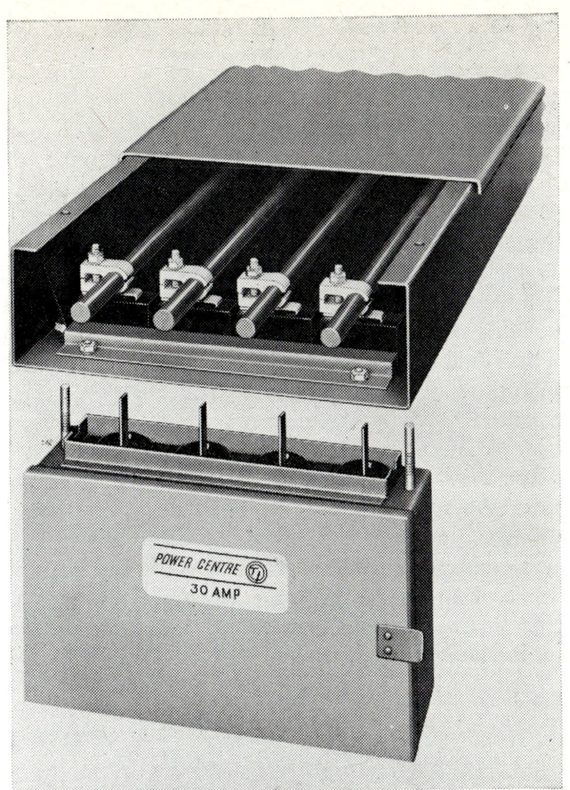

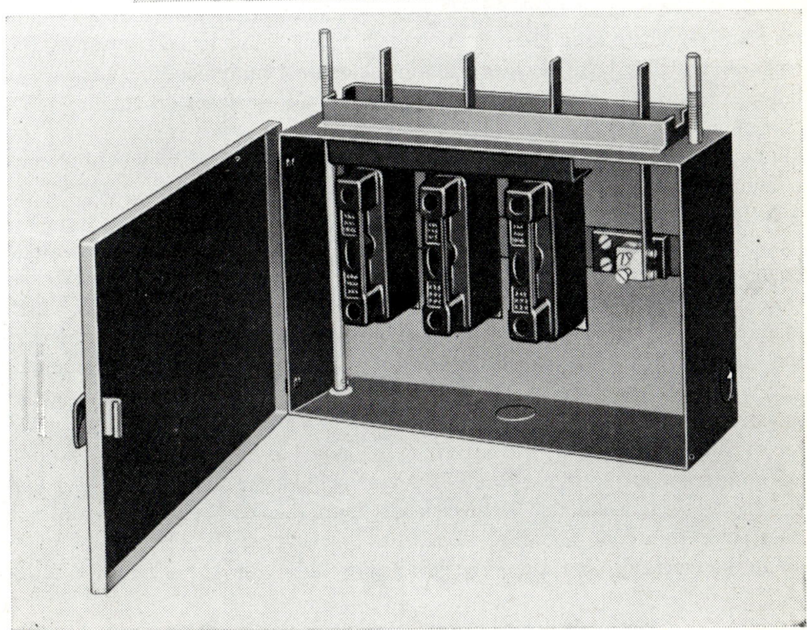

FIG. 32 *Tapping-in boxes for overhead busbar system.*

To facilitate the re-positioning and re-connection of machines, all the control gear should be mounted directly on the machine itself. Thus, all that is necessary to connect the machine to the trunking is to connect a conduit from the tapping-in box to the machine, with possibly the final connection to the box being made with flexible conduit, again to facilitate re-connection of the machines.

The individual busbar sections may be controlled from a busbar panel, the rating of the switchgear and panel being dependent upon the rating of the sections, although very high-rated sections may be controlled from their own circuit-breaker. Where the former method is employed, fuse-switches, with h.b.c. fuses would be mounted on the panel, from which possibly p.v.c.-insulated and armoured cables would be taken to the respective sections, although other types of main cable might be preferred.

In some instances, where the machines connected to the busbar sections constitute the major part of the load, it may be more economical to install a package substation at approximately the centre of gravity of the electrical load. Where space is limited, the substation may be installed on a suspended platform, employing dry-type transformers. This is the most economical distribution method as the secondary distribution cables would be installed for a minimum length of run. It might not be convenient, however, to install the substation inside the factory, in which case the busbar panels would be served from their own circuit-breaker inside the main substation.

42. *Describe the advantages of a lighting trunking system, and state how it is constructed and installed.*

This system is now used extensively in commercial and industrial lighting installations where long lines of fluorescent fittings have to be suspended, the trunking not only being used to contain the control equipment and the wiring, but also to suspend the fittings. The system therefore reduces the conduit work to a minimum and enables lighting fittings to be removed or replaced with the minimum of effort. Also, to a limited degree, the trunking may be used to contain a cable-tap system for serving power appliances.

In the type shown in Fig. 33 the 100 mm by 75 mm zinc-coated, sheet-steel trunking is constructed primarily for carrying fluorescent fittings, knockouts being provided every 300 mm for mounting the fitting end bracket assemblies, so enabling fittings of various lengths to be mounted. The lighting assemblies, which may be suitable for single or twin fittings, are secured to the trunking by bushes, the wiring being taken through the bushes to the control gear which is mounted on a tray inside the trucking. The reflectors are made in two or four parts and are attached to the end brackets by captive nuts. This enables the reflector to be removed for cleaning without disturbing the lamps or assemblies.

The effective length of each section, which includes the coupling sleeve, is 3 m and the trunking is suitable for suspension where the span between suspensions does not exceed 6 m. Standard suspension brackets may be used at any place along the trunking, and are designed to allow the cover to be removed fairly simply. They also include a central hole designed to accommodate a 10 mm hook bolt, so enabling the entire length of trunking to be suspended from the cross-members.

FIG. 33 *Lighting trunking system.*

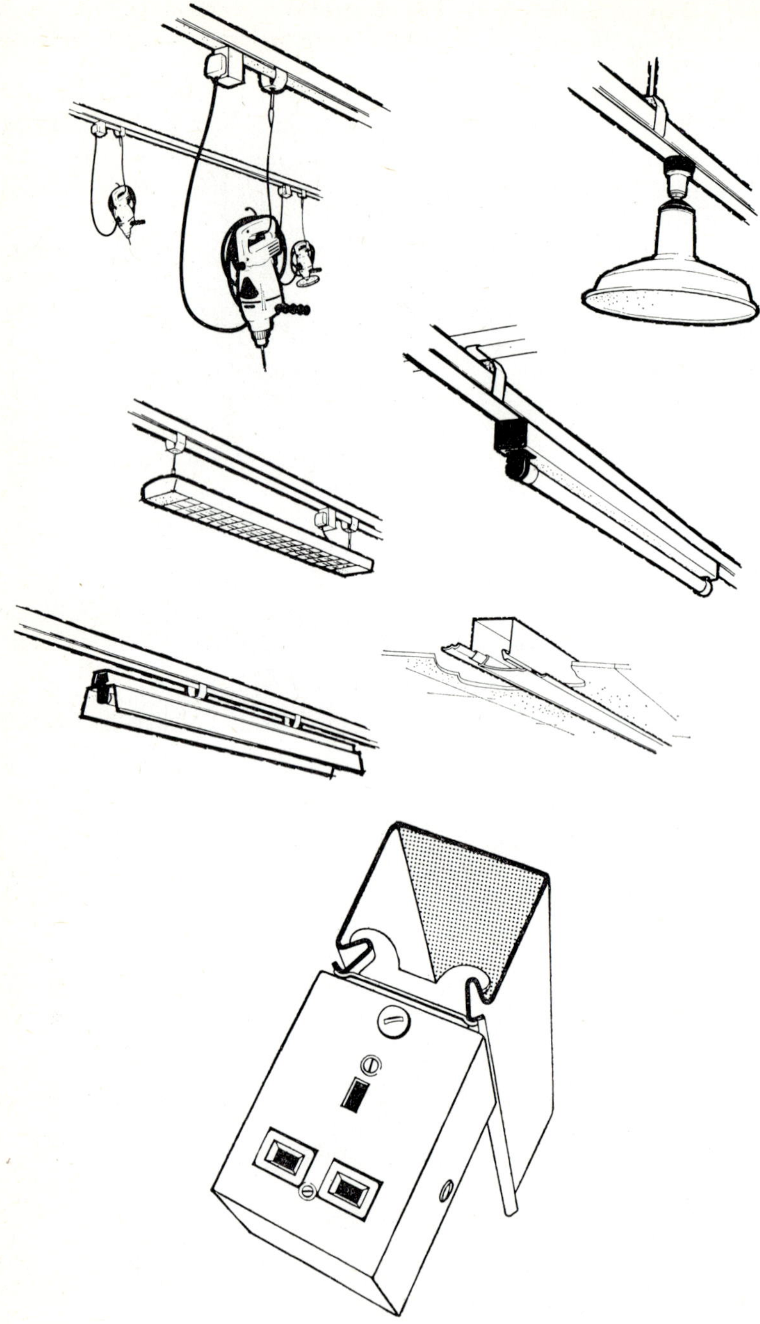

FIG. 33 (*continued*)

Although the control gear is contained within the trunking, the cross-sectional area is such as to allow liberal wiring space below and at one side of the control gear. To facilitate the use of this space, 25 mm knock-outs are provided on each side of the sleeve coupling in order that entries may be made into the trunking.

Where the system is also to be used as a cable-tap system for distributing power to other appliances, etc., cables with a maximum size of 50 mm^2 may be laid in the trunking, tap-off points being provided by all-insulated connector blocks fitted inside the trunking, to which the main cables and the subsidiary cables are attached. Tapping-in boxes containing fuses with a maximum capacity of 30 A, and also a neutral link may then be coupled to the underside of the trunking at convenient points, a 20 mm bush being fitted to the box for this purpose. A locknut on the inside of the trunking thus secures the box to the trunking. Because of the different method of mounting, the box is not interchangeable with boxes of the normal cable-tap system.

Cables with a maximum size equivalent to 4·0 mm^2 may then be used for connecting the fuses and link to the connection box, while conduit or other systems of wiring may be used for connecting the appliance to the tapping-in box. The two systems combined, therefore, give quite a high degree of flexibility.

43. *Describe the types of situation suitable for a floor trunking system and state how such a system is installed.*

This trunking system (Fig. 34) is often used in commercial and industrial installations where remote island positions have to be served from a run installed in the most convenient position. If necessary 20 mm or 25 mm knock-outs may be provided from which conduits are taken to the required location. Alternatively the holes may be drilled on site to suit the actual installation conditions. Conduit terminations finishing above the concrete should be blanked off to prevent the entry of foreign matter into the conduit.

This type of trunking is also used for connecting underfloor heating cables to the supply. By combining the system with skirting trunking, a very flexible system is obtained.

Where segregation of cables is necessary, dividing fillets may be provided, which means that special boxes with cross-over bridges must be fitted at intersection points to ensure the cables are kept segregated.

Individual lengths are joined together by butt straps, and to ensure good continuity, the inside surfaces of the straps are left unpainted together with a corresponding area of trunking. Also, if necessary, the trunking may be of a stepped construction, with either two or three compartments to segregate the cables.

To install, lengths of trunking, normally 3 m long, are laid in position and connected together by butt straps, one end being blanked off and possibly a rising elbow being fitted at the other end to enable the trunking to be connected to surface trunking. Care should be taken to ensure that the trunking is level and

FIG. 34 *Floor trunking system.*

that when the cover is fitted it is level with the floor surface. The conduits are then fitted and the front cover replaced before the concrete is applied. This is to prevent distortion due to excessive pressure against the side of the trunking. Also, as with all sub-floor trunking systems, the concrete should not be laid too tightly against the side of the trunking. When all the cables have been installed and connected, the trunking may be finished by resting chequer plates on a rubber-bonded gasket, the plates being level with the floor surface. Alternatively, a fill-tray cover may be installed, incorporating brass spacing bushes to prevent undue tension on the cover.

44. *Describe the advantages of the under-floor trunking system and state how such a system is installed.*

This system is seen to its best advantage in large commercial installations where the frequent re-positioning of equipment also requires the alteration of light and power outlets. The installation of such a system minimises the necessity of running conduits overhead or across large stretches of the floor and so increases the flexibility and the finished appearance of the installation.

The system is installed in a somewhat similar fashion to the floor trunking system, but to provide the highest degree of flexibility, the trunking is normally laid in a grid formation (Fig. 35). The method also varies from the floor system in

FIG. 35 *Underfloor trunking system.*
(*Courtesy Walsall Conduits Ltd.*)

that both the trunking and top covers are completely buried in the floor. In the type shown in Fig. 35, however, outlet bosses with an 40·0 mm internal thread are attached to the trunking at the required positions. Where the outlet positions are known prior to the trunking being installed, the outlet bushes may be attached as an integral part of the system. If the positions are unknown, however, after the trunking has been laid in position 45 mm diameter holes may be cut in the top surface of the trunking into which the outlet bushes can be fitted. Where the bushes are an integral part of the trunking, stopping plugs are fitted into the outlet. A screw of just sufficient length to finish flush with the completed floor is then screwed into the plug, so providing a marking point for the outlet position. All that is necessary to connect an accessory to the trunking is to remove the concrete screed around the screw, just sufficient to allow the stopping plug to be screwed out and the accessory to be screwed into the outlet.

Although single-compartment trunking may be installed, it is usual to find, especially in large commercial buildings, two or even three-compartment trunking so that cables may be segregated. In such situations, it is usual to find the outlets symmetrically spaced as shown in Fig. 36.

As an alternative to the method of marking the outlet positions by screws, the outlets themselves may be extended to finish flush with the floor surface. The outlets are thus available for immediate use by removing the stopper plug and screwing the necessary accessory into position. Where this type of outlet is used, it is essential to know the thickness of the screed above the top surface of the trunking in order that the correct length of outlet is installed.

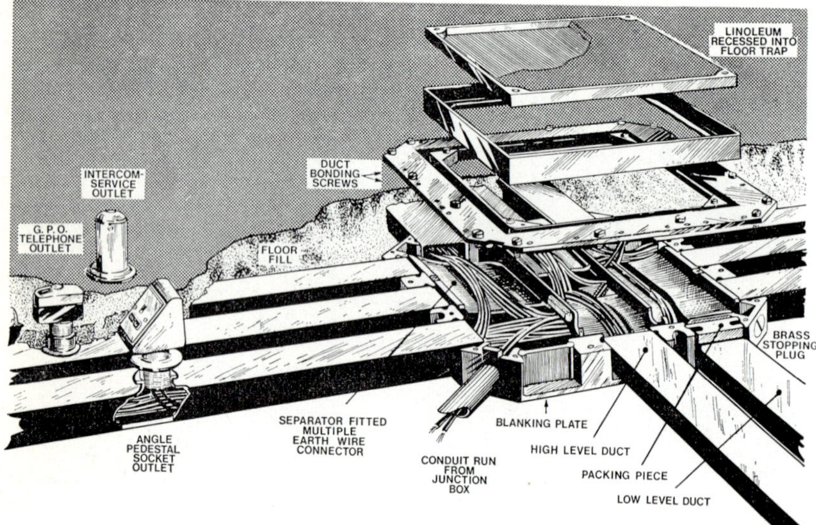

FIG. 36 *Junction box for underflooring systems.*
(Courtesy Walsall Conduits Ltd.)

Where it is necessary to segregate cables cross-over bridges must be fitted in junction boxes, but as this reduces the capacity of the box, if the trunking is to carry its optimum number of cables, a larger box may have to be installed. Junction boxes (Fig. 36) are also provided with levelling screws so that when the covers are fitted they finish flush with the floor surface. In order to ensure sound earth continuity of the trunking system, special couplers are necessary to connect the trunking to the junction boxes.

The cover of the boxes may consist of a pressed sheet-steel tray filled with materials similar to the floor finish, or alternatively, a chequer-plate cover may be used.

Another method of using underfloor trunking is where it is installed in corridors to facilitate services to adjoining rooms for mains or submains, or for serving lighting on the floor beneath the trunking (Fig. 37). The trunking is left un-drilled, all the conduits being arranged to terminate at the junction boxes.

FIG. 37 (a) *Applications of an underfloor trunking system.*

FIG. 37 (b)

45. *What types of skirting trunking are available and how would each type be installed?*

The available types are illustrated in Fig. 38(a), (b), (c) and (d). The type shown in (a) is designed for installation after the wall surfaces have been finished. It is constructed in zinc-coated sheet-steel normally in 3 m lengths, one fillet being incorporated which enables cables to be segregated where necessary, which also serves for the attachment of the front cover. If further segregation is necessary, additional fillets may be fitted. Joints are made by a butt strap at the back of the

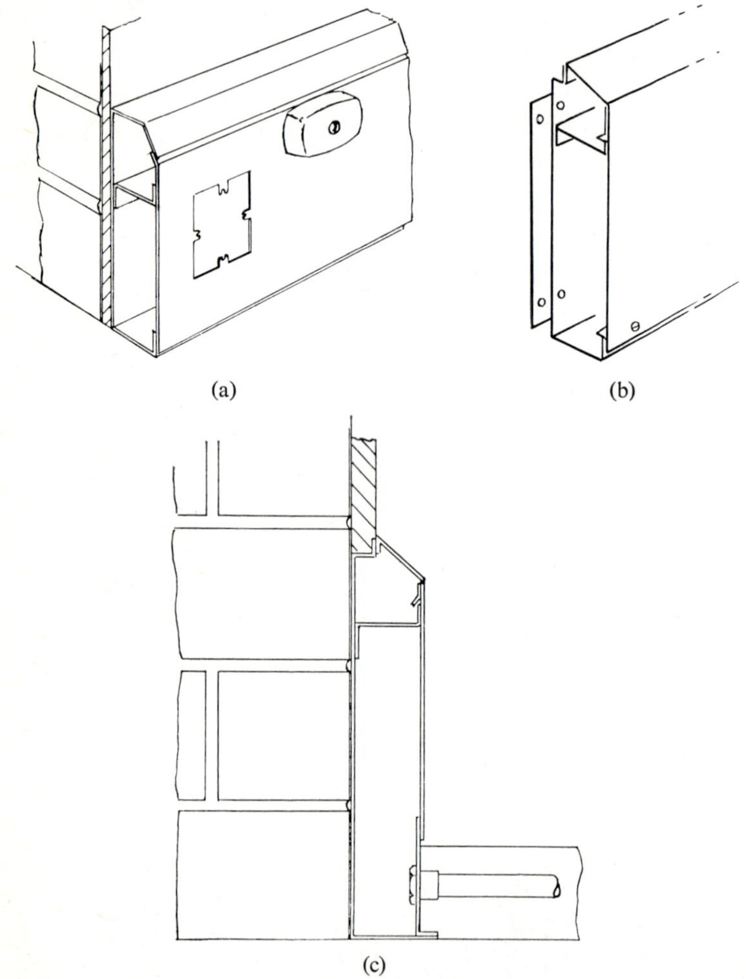

FIG. 38 *Types of skirting trunking system.*

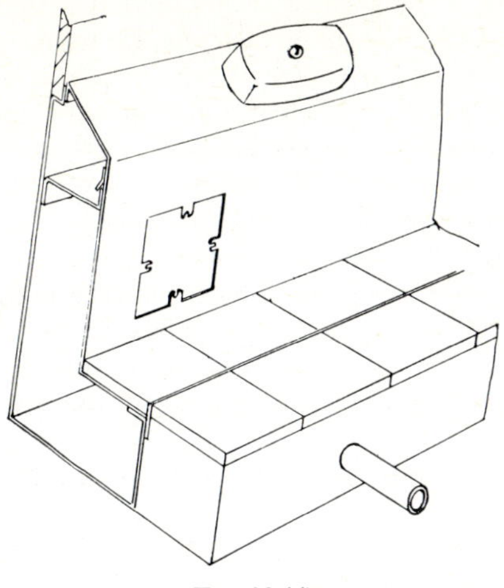

Fig. 38 (d)

trunking, the inner surface of the strip and a corresponding area of trunking being left unpainted to provide good earth continuity. A copper earthing strap may also be provided. Special pieces are necessary at corners and may also incorporate sliding sleeves so that adjustments may be made if the accuracy of cutting is at fault.

Although the trunking is obtainable with the front cover undrilled, it is normally more convenient to have the holes for socket-outlets, telephone connection blocks, etc., drilled in the cover by the manufacturers, so simplifying the erection of the system on site.

The type shown in (b) is similar to that shown in (a) but is designed to be installed before the walls are plastered. The trunking is fastened to the wall, a channel being formed between the wall and the top of the trunking which is intended to receive the lower extremity of the wall plaster.

The type shown in (c) is similar to that shown in (b) except that the trunking extends below the surface of the floor. This type is very suitable where conduits have to be run in the concrete screed, or for making connections to underfloor heating cables. If specified, therefore, knock-outs may be provided suitable for 20 mm or 25 mm conduits.

A further type is shown in (d). This is very similar to the one shown in (c) except that additional capacity is provided in the sub-floor portion. It can, therefore, accommodate a required number of main cables, or alternatively, greater facility is provided for connecting underfloor heating cables. The greater floor space also simplifies the withdrawal and replacement of underfloor cables where necessary. Removable bridge pieces, located every 450 mm, serve not only for fixing the front cover, but act as struts between the front and back portions of the trunking and prevent distortion when the screed is being laid. The front cover of the trunking is designed to enclose both wall and floor parts simultaneously. With the cover in position, therefore, the trunking may be cut to any desired length.

Installation. For type (a) suitable fixings are inserted in the wall which is then plastered. The trunking is cut to the required lengths and fitted to the wall, corner and bridge pieces being fitted where necessary. Where doorways intervene, the trunking may be interconnected by short lengths of underfloor trunking. It is necessary to install this before the concrete screed is applied and any open ends should be blanked off to prevent foreign matter from entering the trunking.

Type (b) is fixed directly to the wall with the bridge and corner pieces in position, so that when the wall is finally plastered, there is a depth of approximately 15 mm of plaster in the recess between the top of the trunking and the wall.

Both types (c) and (d) must be fastened in position before the floor screed is laid and although it is preferable to have the front cover in position while the floor is being laid, before the plastering is commenced, the front covers should be removed to allow the plasterer to trowel off to the front edge of the main trunking body.

Although bridge pieces are inserted to strengthen the trunking, where types (c) and (d) are being installed, care should be taken to ensure that excessive pressure is not applied to the trunking, particularly by expansion of the floor after laying. To minimise the pressure, finishes liable to expand should not be laid too closely to the trunking.

These types of system are now used fairly extensively in domestic and commercial installations, being most advantageous when the points to be served are adjacent to the trunking. The trunking replaces the skirting board normally fitted.

5

Wiring systems in special situations

Temporary installations on construction sites—damp situations—flammable and explosive situations—caravan installations—agricultural holdings

46. *Describe the precautions that must be taken when erecting temporary installations on construction sites.*

Temporary installation should conform with the requirements of the I.E.E. Regulations and Codes of Practice 321, and, therefore, temporary installations must comply with the requirements for permanent installation and certain other precautions. One relaxation, however, is that the requirements for supports in permanent installations need not apply, provided the supports are so arranged that there is no appreciable strain on any cable termination or joint.

Additional safety may also be provided by adopting any of the following methods:

(1) The installations must be in the charge of a competent person who must accept full responsibility for the safety of the installation, its use and for any alteration or extension, and whose name and designation must be prominently displayed close to the main switch or circuit-breaker.

(2) The installation must be inspected and tested in accordance to the requirements of the I.E.E. Regulations at intervals of every three months or less depending upon the nature of the installation.

(3) Supplies should be from double-wound transformers with a secondary voltage not exceeding 110 V r.m.s. single-phase (see Fig. 53, pg. 112), with the secondary winding centre-tapped so that the maximum voltage to earth does not exceed 55 V; or a secondary voltage not exceeding 110 V r.m.s. three-phase, the neutral point being connected to earth, so that the maximum voltage to earth does not exceed 64 V.

(4) Where reduced voltage supplies are impracticable, monitored earthing devices or other devices such as current-operated earth-leakage circuit-breakers may be used for comparable protection.

(5) Double-pole switches or linked circuit-breakers should be used to control every two-wire circuit.

In addition, Code of Practice 321 recommends:

(6) Sufficient socket-outlets should be provided for electrically hand-operated appliances.

(7) Even where not exposed to weather, it is desirable that switches, socket-outlets and other accessories should be of a weatherproof construction.

(8) Transformers should be of the air-cooled pattern.

(9) 25 V handlamps should be used with a maximum of 12·5 V to earth between any live conductor and earth.

(10) Switches controlling 110 V and 25 V supplies should be of the double-pole type and fuses must be fitted in each live conductor.

(11) Tough-rubber-sheathed or p.v.c.-sheathed cables incorporating an earth continuity conductor should be used throughout any temporary installation.

(12) All-insulated lampholders capable of withstanding rough usage should be fitted. Lampholders with pin contacts intended to be pressed into sheathed cables must not be used. Where necessary, lampholders should be of the totally-enclosed pattern to prevent unauthorised use of lampholder adaptors. This is very important where appliances such as electric drills which require earthing are to be connected to the supply.

(13) Plugs and socket-outlets should be constructed to withstand rough usage and preferably be of resilient rubber or synthetic rubber compound. Socket-outlets should be marked with the correct voltage and should not be interchangeable with those used on other supply voltages.

(14) In view of the possible damage to the existing installation (accessories, switches, etc.) temporary installations should not be connected to permanent installation until they have been completed, and in any event, care must be exercised that the temporary installation is not connected before it has been ensured that the permanent installation is capable of carrying the additional load.

The actual installation, however, will vary with the actual site condition. For instance, at the start of a large building site, loose trailing cables in the mud would be inadmissible, and possibly the best arrangement would be for a few high-powered lamps mounted on poles, with possibly socket-outlets at a reduced voltage mounted on the same poles.

Where the site consists of a number of completely erected rooms, it is uneconomic to supply lights and sockets in each room, and therefore, where the lighting circuits are concerned, care must be taken that any loose wires are fastened securely out of reach without imposing strain on the connection.

For small power supplies, a suitable arrangement is to use a unit with the necessary switches, socket-outlets and transformers mounted on it, which could be transported from floor to floor.

47. *Define a damp situation and state the types of wiring system suitable for such situations.*

A damp situation is one in which moisture is either permanently or intermittently present to such an extent as to be likely to impair the effectiveness of an installation conforming to the requirements for ordinary situations. It includes all external situations, communal wash-houses and kitchens, laundries, dairies, cold-storage rooms, boiler-houses and all other indoor situations where water is likely to affect the installation.

The following systems are suitable for damp situations:

Heavy-gauge steel conduit. Where this system is installed, the conduit must be galvanized or sheradized. Stand-off saddles should be installed and all joints must be covered with a good bitumastic paint. Accessories should be limited to the number required to assist the ready renewal and withdrawal of cables. Running couplings should be avoided wherever possible, and all accessories should be provided with rubber gaskets. In extreme cases, it may be necessary to seal inspection boxes with compound.

FIG. 39 *Weatherproof switch.* (*Courtesy Walsall Conduits Ltd.*)

Waterproof switches (Fig. 39) and socket-outlets should be installed where required, and lighting fittings should be of the well-glass type (Fig. 40) for filament lamps or the totally-enclosed type for fluorescent lamps.

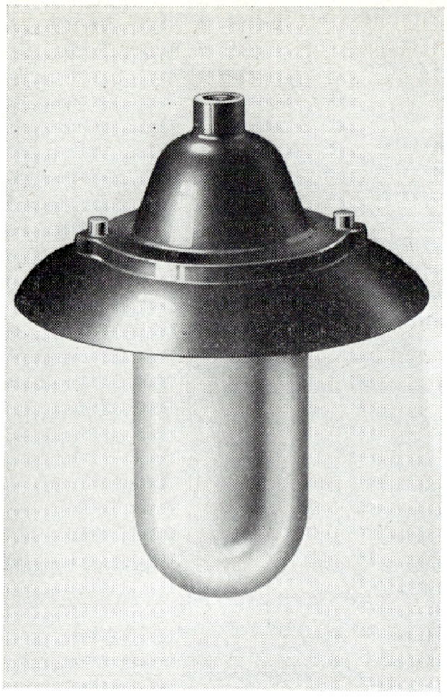

FIG. 40 *Well-glass lighting fitting.*

Mineral-insulated systems. This is possibly the most suitable system when installed correctly. In damp situations, however, it is necessary to install p.v.c.-covered cables, plastic shrouds being fitted at terminations. Weatherproof fittings and accessories are also essential.

Lead-sheathed systems. Although this system may be installed, other systems are normally preferred because of the relatively poor mechanical protection provided by the sheath and the difficulty in maintaining sound earth continuity. Special glands are necessary for maintaining continuity at accessories, which should be weatherproof and as few as possible. Fixing saddles should be constructed of lead or lead-alloy.

Armoured p.v.c.-insulated systems. This is another system which is particularly useful in outdoor and indoor damp situations, provided that the ambient temperature is within the permissible limits. In such situations, an overall p.v.c. covering with plastic shrouds at termination is necessary. No special jointing techniques are necessary and standard boxes can be used. Glands with cone-type clamps are necessary, however, so that when the outer sheath is removed, the armouring wires can be cut to length and securely fastened to the clamp to ensure sound earth continuity, the current-carrying conductors being taken into the box. P.V.C. tapes may be used to replace any of the sheath removed. If the resistance of the steel armouring is too high to satisfy the requirements of the

I.E.E. Regulations, one or two strands of copper may be included to reduce the overall resistance. One advantage of this system is that multi-core cables are available if required which could mean a considerable saving in space and fixing time.

On the other hand where the number of conductors is small and the number of terminations large, the comparable cost is higher than that of other systems, which would probably be preferred.

Whichever system is installed, it is essential that all clips and their fixings are of a corrosion-resistant finish. In damp situations brass or sheradized screws should be used instead of steel screws, although it must not be forgotten, when selecting clips and fixings, the possibility of corrosion occurring when dissimilar metals are installed in corrosive atmospheres.

48. *Describe the essential requirements of flameproof equipment and state in what situations such equipment might not be suitable.*

For any apparatus to have a flameproof certificate it is required that the containing case, or other enclosure, should withstand without injury any explosion of prescribed flammable gas that may occur within it under practical conditions of operation within the rating of the apparatus (and any recognised overloads), and will prevent the transmission of flame such as will ignite any prescribed flammable gas that may be present in the surrounding atmosphere.

FIG. 41 *Flameproof switch. (Courtesy Walsall Conduits Ltd.)*

Flameproof equipment (Fig. 41) is constructed with specially machined faces and wide flanges, and, therefore, the main consideration in design is the size of flanges and the dimensions of the gaps between faces to correspond with the division number of the situation in which the equipment is to be installed. It is important then, when selecting flameproof equipment, that it bears the certificate of registration relating to its type which must correspond to the division number for which it has been tested.

Under normal manufacturing processes it is virtually impossible to make the enclosure completely gas-tight and the entry of small quantities of gas into the enclosure is acceptable provided that the gaps are so designed that any explosion of gas inside the enclosure, caused by operational arcing or failure of the insulation, will not penetrate the gap and ignite any flammable gas in the surrounding atmosphere.

It is clearly important that flameproof equipment should be frequently inspected and tested, and only highly-skilled operatives should be allowed to undertake the necessary maintenance work on such equipment. It is also important that, in addition to the wiring system being flameproof, any associated apparature, including handlamps, must also be of flameproof construction.

Even in those flammable situations where flameproof equipment cannot be guaranteed to give satisfactory protection and therefore other protective devices have to be installed, it is sound practice to install flameproof equipment as an additional precaution.

For the purpose of certification, areas are now classed in divisions as follows:

A division 1 area is defined as an area within which any flammable or explosive substance, whether gas or vapour or volatile liquid is processed, handled or stored and where, during normal operations, any explosive or ignitable concentration is likely to occur in sufficient quantity to produce a hazard.

All the methods of precaution given in question **49** are applicable to this type of situation.

A division 0 area is defined as an area or enclosed space within which any flammable or explosive substance whether gas, vapour or volatile liquid is continuously present in concentration within the lower and upper limits of flammability.

In these areas, flameproof equipment is not a guaranteed method of protection, and intrinsically safe or pressurised equipment (see question **49**) should only be used where it is unavoidable.

A division 2 area is defined as an area within which any flammable or explosive substance, whether gas, vapour or volatile liquid, although processed or stored, is so well under conditions of control that the production (or release) of an explosive or flammable concentration in sufficient quantity to constitute a hazard is only likely under abnormal conditions.

The conditions for division 2 areas are outlined in C.P.1003. Briefly, in these areas, flameproof equipment need not be installed, except that sparking or arcing contacts must be contained in flameproof enclosures, and also the construction of non-flameproof equipment must be mechanically sound and sufficiently strong.

49. *Describe the methods of protection in flammable situations afforded by the following: (a) certified flameproof equipment, (b) certified instrinsically-safe equipment, (c) segregation, (d) pressurization, and (e) ventilation.*

Certified flameproof equipment. See question **48**.

Certified intrinsically-safe equipment (Fig. 42). Where a certificate has been granted for an intrinsically-safe circuit or piece of apparatus, it simply means that there is not sufficient energy in the circuit to ignite any flammable gas or vapour when maximum current is flowing, possibly due to a short-circuit in the wiring. Hence, it cannot be used for ordinary lighting and power circuits, but it

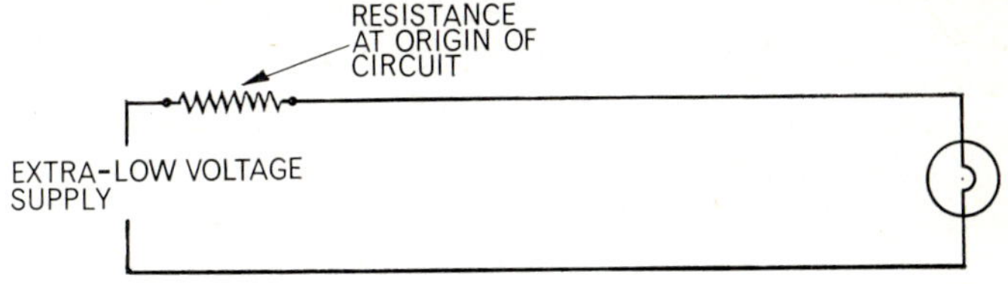

FIG. 42 *Intrinsically safe circuit.*

has a distinct advantage in circuits where low currents are required at low voltages. Thus intrinsically-safe apparatus has valuable applications in mines for magnetic exploders and in hospital operating theatres where low-powered bulbs are used in endoscopic instruments during nose and throat operations. In industrial installation, intrinsically-safe handlamps are found to operate satisfactorily in garage pits, etc., where they are supplied from a 6 V supply with a maximum current of 1 A.

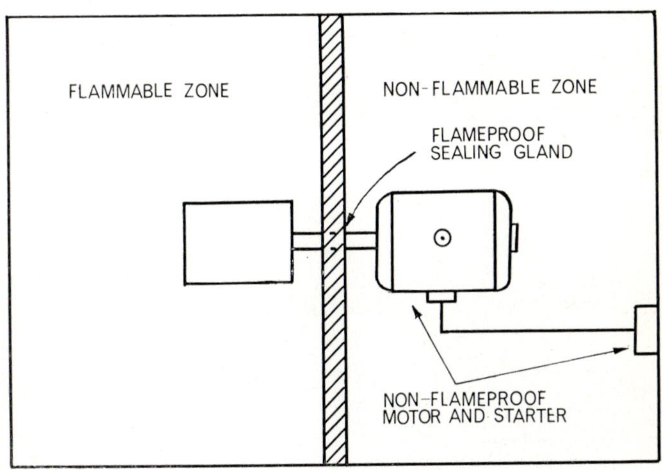

FIG. 43 *Segregation of equipment in a flammable situation.*

In intrinsically-safe circuits, a device is usually included in the circuit to limit the current to a safe value. This is situated at the origin of the circuit, and no attempt must be made to alter its location or vary its value.

Segregation (Fig. 43). Possibly this is the most effective method of preventing fire and explosion hazards, a particular example being where electric motors are installed in oil refineries. The motors are installed in a room separated by a dividing wall from the pump house with no direct means of communication between the two rooms. Specially packed glands are required where the driving shaft passes through the wall, but as the motor room constitutes a non-flammable situation, normal industrial type motors and equipment can be installed. Should the motors have to be controlled from the pump-house, flameproof push-buttons may be installed in the latter, a flameproof stopper box being installed at the junction of the two zones.

Similarly in garages, although the petrol pumps must contain flameproof accessories and be wired in m.i.c.c. cables, the switches may be installed outside the danger zone and need not be a flameproof pattern.

Other applications of segregation include spray booths, where the equipment can be mounted outside the booth, and hospital operating theatres, where the danger zone is in the vicinity of the operating table.

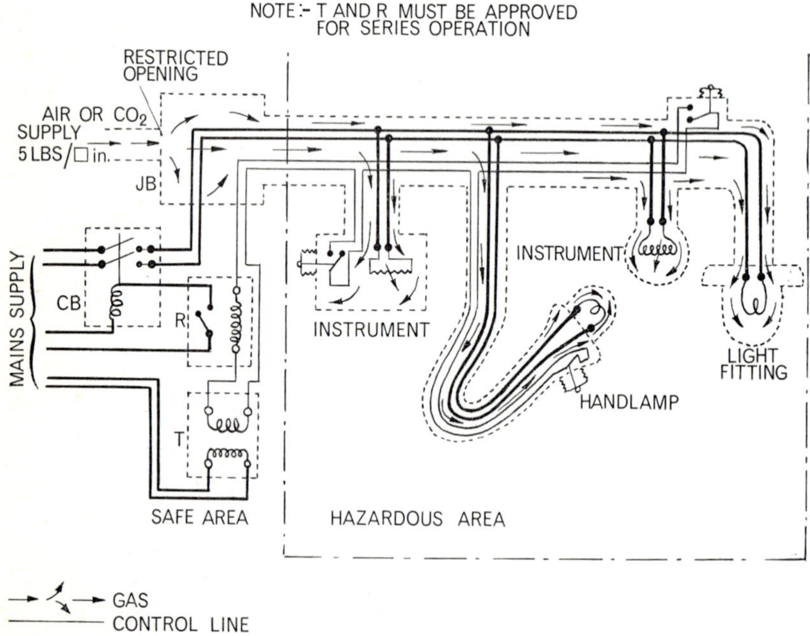

FIG. 44 *Illustrating pressurization of equipment in a flammable situation.*

Pressurization (Fig. 44). Where gases such as hydrogen or acetylene constitute the fire risk, segregation should be adopted where possible, but this is sometimes difficult where lighting is concerned. As certified flameproof equipment cannot be guaranteed adequate protection against such gases, then pressurization should be adopted.

In this system, the interior of the fittings are supplied with fresh air or an inert gas, such as nitrogen, at a slight pressure. The pressure is maintained constantly so that the fitting breathes outward, preventing flammable mixtures from being drawn into the fitting as it cools down. A further precaution is to install flameproof fittings suitable for division 2 areas, which in the event of air failure may still provide a possible safeguard. Indicating instruments should be installed in the danger area, and also pressure controlled devices to ensure continuity of service. In many instances, it is necessary to link the two services together so that in the event of air failure, the electrical equipment in the danger area is disconnected.

It will be appreciated that a system such as this requires more supervision than other types of safeguards, and generally it would only be installed in situations where other systems are unsatisfactory.

Ventilation. Although ventilation itself is not a sufficient safeguard, in many flammable situations, where the risk is small and where, under normal conditions, the concentration of flammable mixture is not sufficient to constitute a fire hazard, an air extraction system may give adequate protection. The extraction system should be linked with the processing system, so that a shut-down of the extraction plant also shuts down any of the electrical equipment associated with the processing plant. Where this is impossible, and concentration of flammable mixtures are possible if the extraction plant fails, then other precautions must be taken. Where ventilation systems are installed, however, they should be designed so that uncomfortable working conditions are not produced, which make personnel tend to switch off the extraction plant while work is in process.

50. *Describe the types of wiring system suitable for flameproof situations and state what precautions are necessary when installing them.*

Conduit systems. It is essential that solid-drawn steel conduit should be used and the following precaution should be observed.

For surface work stand-off saddles should be used to prevent possible corrosion behind the conduit.

Certified flameproof unions should be used where normally running couplings would be installed.

Where fittings and accessories are of a flameproof construction the threads must be at least 25 mm long and all joints must be painted to prevent corrosion.

Locknuts should be provided each side of couplings and also behind accessories and fittings to prevent any possible vibration from loosening the connection and so introducing high resistance in the earth-fault path.

Where a conduit passes from a hazardous area to a safe area, a flameproof box must be installed at the junction of the two zones.

Lead-sheathed steel-armoured cables. This system can be used advantageously where the number of terminations is low and the current to be carried high. The lead sheath may be protected by single-wire, double-wire or double-tape armouring with an overall jute or hessian covering to guard against corrosion. The cable should be well supported by cleats of the all-embracing type. The insulation may be of paper, varnished-cambric, p.v.c. or vulcanised rubber. As flameproof sealing is essential at terminations, the sealing box should be of a type associated with the certified accessory or fitting, but where soldering is not permitted, cone-type glands may be installed.

Mineral-insulated copper-sheathed cables. This is generally the most preferred system for flammable situations, the reason being that the cable itself is fireproof, and if suitably protected, virtually corrosion proof. Installation is similar to that adopted for non-flammable situations except that as flameproof equipment is threaded to accommodate 2·5 mm long threads, special flameproof glands are necessary and additional locknuts are required as in the steel conduit system.

Armoured p.v.c. cables. Where the current to be carried is large, such as in mains or sub-mains, this would possibly be the most economical cable to install, but for wiring systems it would be more economical to install a conduit or m.i.c.c. system. The conductors may be of copper or aluminium and special glands are necessary at fittings and accessories.

Aluminium-sheathed cables. Note, this refers to the sheath only. The smallest cross sectional area for aluminium conductors recognised by the Regulations is 16 mm^2 which precludes it from being used as a wiring system. It can be used advantageously where the current is large, such as for mains and sub-mains, and even for wiring systems if the current-carrying conductors are of copper, but usually, for wiring systems a conduit or m.i.c.c. system would be preferred. Where the conductors are of aluminium, possibly crimping methods would be employed at joints and terminations, and again special flameproof accessories are necessary.

The sheath must be protected against corrosion, which may be by incorporating a p.v.c. covering, but the sheath must also be adequately protected from sparking by mechanical impact which might cause an explosion. It might therefore entail the sheath being protected with a steel-wire armouring with possibly a further p.v.c. covering. One point that must be taken into account when installing aluminium-sheathed cables is that the radius of any bend is far greater than that of a cable of the same size but with a different type of sheath.

Whichever system is installed it is essential that the earth-continuity path is efficient throughout the entire life of the system, hence the reason why the right type of glands and fittings are used, and why locknuts are provided where there is a possibility of vibration occurring.

A point, which is not always observed, is that, as flameproof fittings and accessories are often of considerable weight, fixings must be firm and sufficiently strong to support the weight throughout the life of the installation.

51. *Describe the main points to be observed when installing electrical services in residential and mobile caravans.*

Termination of site wiring. Every termination of the site wiring must contain means of protection against excess current, means of isolation and means of earthing.

If the termination is external and not attached to the caravan, it must be in a weatherproof box and may consist of terminals marked 'LIVE', 'NEUTRAL' and 'EARTH' which may be in the isolator or be in a watt-hour meter (Fig. 45). Or it may be a socket-outlet controlled by a switch or circuit-breaker. The socket-outlet must be of a non-reversible pattern with provision for earthing, of rating 15, 30 or 60 A, and incorporate a means of holding the plug firmly in position.

Where a touring caravan is to be served, an indelible notice must be fixed near the termination, indicating the voltage, whether a.c. or d.c., frequency if a.c., and the maximum permissible load.

Where the site wiring is not attached to the caravan, and consists of terminals (as above), the caravan may be connected by rubber-insulated, p.c.p.-sheathed flexible cord or flexible cable terminating in a connector for the caravan inlet, or by approved cables terminating in the main switch. However, where the site termination takes the form of a socket-outlet, only flexible cords or flexible cables should be used, terminating in a suitable plug, and to a connector suitable for the caravan inlet.

Note that the reason p.c.p. cable is advocated is that it is resistant to the effects of sunlight and retains its flexibility in low temperatures.

Every connector for a caravan inlet shall be of a non-reversible type, with provisions for earthing, include a means of securing the connector in position, be of 15, 30 or 60 A rating and be provided with untouchable contact tubes and a means by which the cable may be securely anchored to the connector.

Caravan installations

Inlet. A single inlet must be provided for all caravans connected by a connector and must be of a non-reversible type with provision for earthing; be of 15, 30 or 60 A rating, as appropriate; include a means of holding the connector firmly in position; be provided with recessed contact pins; be not more that 2 m by a direct route from the caravan main switch; be accessible only from the outside of the caravan; be sited remote from storage vessels or pipes containing water, gas or oil; and be so sited and protected from mechanical damage and other damage from weather and road travel.

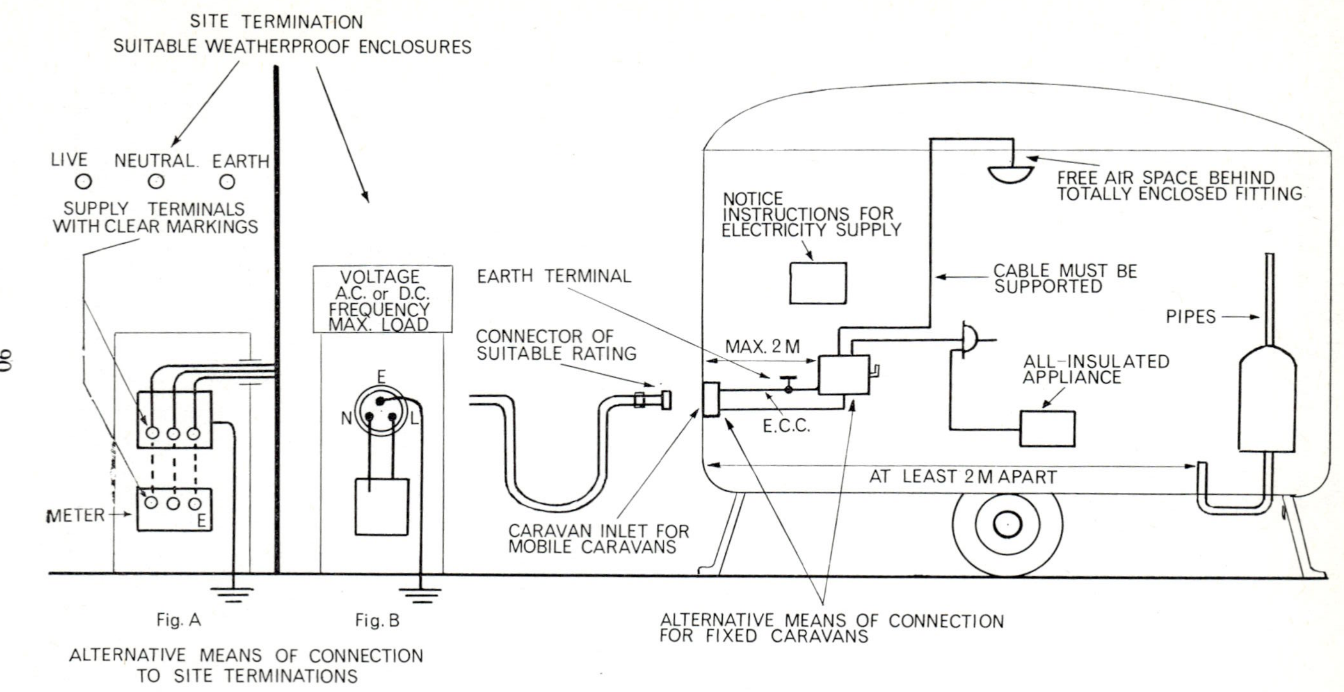

FIG. 45 *Precautions to be taken in caravans and caravan sites.*

In all mobile caravans, a notice of durable material giving instructions for connecting and disconnecting the caravan from the site wiring must be fastened near to the main switch.

Cables. For the fixed wiring inside the caravan, rubber- , p.v.c.- or h.o.f.r.-sheathed cables must be used.

All fixed wiring must be prevented from coming into contact with metalwork of the caravan or any other services by a barrier of non-conducting material, such as spacers of wood or asbestos between the cable and the metalwork, or in inaccessible positions by suitable spacing. Note that if the cable has to pass through metalwork, the cable must be protected by an *insulated* bush. Precaution must also be taken against the effects of heat, and corrosion in damp situations.

Where clips are necessary, they must be of corrosion-resistant material and must be so spaced as to prevent contact of the cable sheath with any metalwork other that that of the electrical installation. Where exposed, cables must be supported at intervals not exceeding 250 mm for horizontal runs and 450 mm for vertical runs, except that in a residential caravan, where the cable passes through joists at approximately 450 mm intervals and is securely bedded in thermal insulation material such as fibre glass, it need not be further fixed.

Lighting fittings. Flexible pendants are not permitted, and the fitting must be mounted directly on the lining or structure of the caravan. Where lighting fittings enclose tungsten-filament lamps, they must be designed or mounted so that air can freely circulate between the fitting and the caravan.

Earthing. All earth-continuity conductors shall preferably be contained in a composite cable, and, in any event, must be insulated from the structural metalwork. Where a separate earth-continuity conductor is installed, it must be not less than 2·5 mm^2 in cross-section and must be insulated. Any earth-continuity conductor must terminate at an earthing terminal insulated from structural metalwork and placed near to the main switch. The terminal must be connected to the earthing-pin of the caravan inlet for a mobile caravan and to the means of earthing provided on site for a residential caravan. It is, however, preferable to use all-insulated or double-insulated apparatus wherever possible.

Precautions near a fixed bath. Where electrical apparatus with exposed metalwork is contained in the same compartment as a fixed bath or shower which can be in metallic connection with earth, such as through water pipes, the exposed metalwork of the electrical apparatus and the bath or shower must be bonded together and earthed in accordance with the Regulations on earthing, unless the electrical apparatus is enclosed in non-conducting material, such as in a cupboard, and in such a manner that it cannot be touched by anyone using the bath or shower.

Testing. Every installation must be tested in accordance with the Regulations and it is recommended that mobile caravans are tested every twelve months. Where a caravan is disconnected for a period longer than three months, it must be re-tested before being connected to the supply.

General. In addition to the above requirements, any caravan installation must conform with the requirements of the general I.E.E. Regulations where they are applicable. Fig. 45 illustrates many of the precautions to be taken in caravan installations.

52. *Describe how lighting, power and heating services may be installed in an agricultural holding.*

Main switchgear. The main switchgear must comply with the general requirements of the Regulations and in addition it must not be installed within reach of livestock; or be in any position where access may be impeded to it by livestock, account being taken in the event of panic by livestock; where more than one building, including glasshouses, is to be served a separate means of control must be installed in or adjacent to each building; where access is possible to live parts in which different voltages exist, a notice must indicate such voltages; every point, *including socket-outlets* must be controlled by a switch; for motors driving machinery, the *on* and *off* positions must be clearly indicated.

It must be remembered that, in addition to the above, the switchgear must be designed and installed to withstand the adverse conditions which are often met in agricultural situations. These involve risk from mechanical damage, corrosion due to damp, steam, ammonia fumes, lactic acid and milk fats, coke and oil fumes, sulphur fumes, and possibly high temperatures.

Cables. Cables sheathed with rubber must be used only in dry, clean situations indoors.

Non-served metal-sheathed paper insulated or varnished-cambric-insulated cables can be used only for dry, clean indoors situations and for clean outdoors situations.

Bright-wire-armoured cables must not be used outdoors.

Non-sheathed twisted-twin and parallel-twin flexible cords must not be used.

All cables must be placed out of reach of livestock and clear of all vehicles.

Cables sheathed with p.c.p or having an h.o.f.r. sheath must not come in contact with liquid creosote.

Where additional protection against mechanical damage is necessary, it must, wherever possible, be provided by the use of non-metallic materials, for example, non-metallic conduit, and in any event light-gauge metal conduit or heavy-gauge metal conduit must not be used.

Long runs of cable along the sides of buildings must, wherever practicable, be on the outside of the building and as high as possible.

Care must be taken that discharge-lamp cables are not adversely affected by their surroundings.

Considering the situations in which cables are to be installed, p.c.p.- or h.o.f.r.-sheathed cables would be suitable for barns, general farmstead buildings, stables, cowsheds or piggeries, light buildings, drying sheds, Dutch barns and outdoor situations. Where the temperature exceeds 115° C, either h.o.f.r.-sheathed or mineral-insulated cables would possibly be used. In dairies, where movement is restricted and the conditions particularly adverse, the cables should be enclosed in heavy-gauge galvanized steel conduit. In glasshouses, because of the possibility of corrosion from spraying liquids, lead-covered or p.v.c.-sheathed-mineral-insulated cables should be used, lead cables being painted with corrosion resistant paint.

Where p.c.p.- or h.o.f.r.-sheathed cables are installed, normally without protection, the following precautions should be taken:

For single-phase supplies, the cable should be twin-cored, incorporating an earth-continuity conductor, even where the cable is enclosed in metallic conduit.

Where protection against mechanical damage is necessary, runs should be sited so that the building structure protects the cables. Thus, cables should, if possible be at high level, out of reach of moving machinery, ladders, etc. In vertical runs, advantage should be taken of running the cables in corners or behind architraves. Where protection against mechanical damage has to be provided, it may be enclosed in tubing or channelling of hard insulating material (not metal).

It is essential that wiring and accessories are well out of reach of livestock which are very susceptible to electric shock.

Cables must be supported as in normal industrial situations and must not come into contact with the metalwork of other services, etc. The support should be corrosion-resisting clips or saddles, or cleats of porcelain or corrosion-resisting metal, so designed as not to damage the cables.

Where passing through walls, floors, etc., cables should be enclosed in bushed non-absorbent incombustible conduit or protected by hardwood casing. If the cables have to pass through metalwork, the hole must be bushed.

Where practicable, all-insulated fittings and switches of a robust construction should be used and, where necessary, they should be dust-proof and water-tight. Where subject to mechanical damage, they should be of metal which must be earthed. The number of switches should be kept to a minimum commensurate with economy and convenience, and sited at high levels, well out of reach of livestock, but convenient for ready operation by farm workers.

Wiring between buildings. Normally, in most agricultural situations, it is possible to serve individual buildings by overhead wiring, and where the span is less than 30 m, p.c.p- or h.o.f.r.-sheathed cables may be mounted on insulators, but where the span is over 30 m they should be supported by a catenary wire. In each case, the height of the span must be not less than the permissible limit (Table B7m of the Regulations).

Where cables are installed on exterior walls, they should be run as high and straight as practicable, using clips or cleats with brass or sheradized fixing screws. Lead-in tappings must be situated so that the shortest run to the main switchgear is provided, the cables entering through the building in an insulated tube, a drip-loop being provided in the cable outside the building. In some instances, the overhead lines may terminate outside the building, in which case a composite cable should be used between the cables and the switchgear.

Where poles have to used, such as in long runs or where outdoor lighting is necessary, they should be of red birch or larch and of sufficient length to enable them to be sunk to a sufficient depth and yet maintain a minimum height of span not less than 5 m.

Even for spans less than 3 m, conduit is not permissible, and where cables have to be buried underground, the depth should normally be not less than 600 mm and in, some instances, 1 m where drainage has to be considered. All stones should be removed from the trench, the cable then being covered to a depth of 75 mm with riddle earth. Cable covers are then laid on top of the earth, and where tiles are used, they should be fully interlocked and should follow cables accurately around bends. Under no circumstances should the cables be left under tension.

Where glasshouses are in banks or rows, the ridge may be used to support the cables provided that the height of span is permissible. Tappings to the individual glasshouses should be made in watertight joint boxes.

Lighting fittings and accessories. In normal dry situations, free from excessive dust or fumes, non-watertight fittings may be used, but where considerable moisture or fumes are present, watertight fittings should be employed. In very dusty situations such as those used for grinding, food-mixing, etc., dustproof or watertight fittings must be installed.

In general, pearl gas-filled lamps would be used, although fluorescent lighting would possibly be preferred in certain parts of the installation.

Fittings should give adequate illumination without glare and should be protected where liable to mechanical damage, anti-vibration devices being fitted where excessive vibration is likely.

If possible, fittings should be mounted directly on the building structure but where pendants are necessary, circular sheathed flexible cords must be used. Where swinging is possible, conduit should be used for long suspension with adequate drainage at the lower end and the fitting sealed to prevent the entry of moisture.

Horticultural holdings, apart from offices, should be regarded as damp situations, and watertight fittings should be installed.

Handlamps should be of the all-insulated pattern and long flexible cords should be avoided by installing sufficient socket-outlets. The latter must be switch-controlled and no switch should be contained in the handlamp. Where low-loaded accessories also have to be installed it is preferable to serve them from a double-wound transformer, the centre-point of the secondary winding being connected to earth, so that the voltage to earth does not exceed 25 V and in any event must not exceed 50 V. A portable transformer may be used, served from a 240 V socket-outlet with the necessary facilities for earthing.

Where the installation is not of the 'all-insulated' construction, and damp conditions prevail or earthed metalwork is present, the voltage should be reduced even further, either from a 12 V supply or a 25 V double-wound transformer with its centre-point connected to earth.

All Edison-type screw lampholders must be shrouded in insulating materials, be suitable to withstand heat, and be of the drip-proof type.

Boxes and fittings used with twin-sheathed cables should be of insulating material fitted with cable gripping glands. In exposed situations, they should be watertight and, where exposed to mechanical damage, be of metal which must be earthed. Boxes should preferably be sealed with compound.

Where cables are installed in heavy-gauge galvanized conduit, such as in the dairy, boxes and fittings must be of a similar construction. Conduits should be kept clear of the wall by using stand-off saddles and, generally, the installation should conform to the requirements for damp situations.

Although watertight socket-outlets are necessary in damp situations, for other locations, robust, industrial-type, switch-controlled socket-outlets may be fitted.

Motive power. The construction of motors and control gear will depend upon the situation in which they are to be installed. Generally, protected-type motors are suitable where dust, fumes or damp are not present; totally enclosed for more arduous situations, weatherproof for damp situations, dust-proof where excessive dust is present, and flameproof motors for flammable situations. Single-phase motors are found more often in farm installations than comparable industrial installations, and where a single-phase three-wire distribution system is used, medium-voltage, single-phase motors may be installed.

Motors may be fixed or transportable, mounted on light bogies or platforms, the transportable type often being used to drive different machines. Transportable motors should be of the totally-enclosed surface-cooled type. The control equipment should be weatherproof and mounted on the same bogie, and so mounted that ready access is provided for the controls. Single-phase transportable motors should be connected by a three-core cable terminating in a plug. A drum or bracket should be fitted for storing the cable when it is not in use. Robust socket-outlets are necessary, the industrial type for indoor use and the watertight type for outdoor use.

The installation of control gear should comply with the general Regulations, but the construction is again dependent upon the situation in which the gear is to be installed. It should, however, be as simple and robust as practicable. Direct-on-line starting would be used for most motors not exceeding 3·75 kW, but for larger sizes some form of current limitation at start is normally necessary.

Where the motor is fixed, it may form an integral part of the machine or may be mounted separately. In the former method, the final sub-circuit should terminate at the control gear and may be a p.c.p.-sheathed cable or other recognised form of cable. Where the motor is mounted separately, the control gear should be mounted on a wall or stanchion adjacent to the motor, and if the latter is mounted on slide-rails, a three-core flexible cable would be necessary for single-phase motors and a four-core cable for three-phase motors.

Usually Vee-rope drives would be used, although flat-belt drives may be used where suitable.

Heating. The heating services can conveniently be divided into the following: water-heating, steam raising and space heating, but whichever service is being installed, due regard must be given to the fire risk involved.

Where high-temperature surface appliances are installed, there must be no risks of surrounding surfaces being scorched or ignited, and guards must be provided to prevent accidental contact with flammable materials.

Heat-resisting flexible cords must be used and luminous indicators must be provided to indicate when the circuit is live. All heaters must be accessible for maintenance purposes and if of a non-insulated construction must be efficiently earthed.

Water heating. Probably the most satisfactory method of water heating is to install thermostatically-controlled pressure water-heaters, so enabling several points to be served, although for very long runs, it might be advisable to install local heaters of the free-outlet or instantaneous type.

Steam raising. Where easily dismantled appliances have to be sterilised, small self-contained chests incorporating a small electrical steam generator may be installed. For sterilizing *in situ*, that is milking plant, etc., separate steam raisers with a higher evaporative capacity are more satisfactory. If they can be arranged to operate on the thermal-storage principle, a further advantage is that they may operate on an off-peak tariff.

Space heating. Space heating is used extensively in agricultural holdings. In situations where heating is only necessary for short periods, unit or radiant heaters, preferably at high level, may be installed. For crop drying, fans and heaters in trunking are satisfactory, but dust-proof switchgear is essential if grain dust is likely to be contained in the atmosphere.

Space heating is also required in piggeries and poultry houses. Where possible heating cables may be buried in concrete floors, use being made again of off-peak tariffs. Night-storage heaters may be installed on the same tariff. Alternatively, low-powered radiant heaters may be installed at high level in piggeries, and tubular or convector heaters in poultry sheds. A storage heater may also be required for piping warm water for drinking purposes. Specially heated cabinets are also required for incubating which may be served from watertight switch-controlled socket-outlets.

In glasshouses, non-luminous heaters with a watertight, rust-proof finish should be installed, and, if soil warming is required, it is preferable to install bare conductors at extra-low-voltage served from a double-wound weather- and rust-proof transformer, In some instances, the one transformer may be made transportable to serve several glasshouses.

Earthing. In situations accessible to livestock, the installation must be, as far as practicable, of the all-insulated pattern, but, even so, an earthing terminal connected to the earth-continuity conductor must be provided at every outlet.

In situations accessible to livestock, steel conduit must not be used as a sole earth-continuity conductor, although it may be used to supplement a separate earth-continuity conductor (normally contained in the conduit.) The earthing lead must be adequately protected against disturbance or damage by livestock or passing mechanical implements. Where the earthing lead is not enclosed in conduit or equivalent mechanical protection, it must be of the armoured type. Because of the overhead method of distribution, earth-electrode systems are used frequently in agricultural holdings, and as cattle are extremely susceptible to shocks, even at extra-low-voltage, the electrode must be buried or situated where it cannot possibly be touched by livestock. Although buried, the earth lead should be as short as is practicable.

Conduit used for protecting sheathed cables, metal suspension chains for lighting fittings and lighting fittings containing metal filament lamps, which are exempt from earthing in industrial installations, must be earthed in farms, unless the metalwork is well out of reach of livestock and accidental contact by passing machinery.

Every switch or other means of control or adjustment must be out of reach of wash troughs, sterilising equipment or any other extraneous bonded and earthed metalwork.

6

Earthing

System earthing—methods of protection against dangerous earth-leakage currents—P.M.E.—direct earthing—earth-leakage circuit-breakers—earth-electrode systems.

53. *Describe the purpose of earthing the non-current carrying metalwork of an electrical installation.*

The purpose of earthing (and bonding) non-current carrying metalwork is to minimise the danger to life and property. Apart from foolhardy tampering with live conductors, shock risk arises when an earth fault occurs on an installation and raises the potential of the non-current carrying metalwork to a dangerous value with respect to the general mass of earth.

Although fire risks may be attributable to other electrical faults, undoubtedly fires are often caused by faults in the earth-continuity conductor, such as high resistance which renders the protective device inoperative. It is essential, therefore, that protection against the possibility of electric shock and fire risk is effective.

To this end, the protective device must be able to detect all fault conditions, and must be capable of isolating the faulty section before any dangerous conditions can develop. While the device must be sensitive enough to protect that part of the plant with which it is associated, it must not affect any healthy circuit, nor must it operate unless the magnitude of the earth-leakage current is such that danger actually exists.

54. *Describe the means which may be adopted for protecting installations against dangerous earth-leakage currents where protection is provided by fuses or excess-current circuit-breakers.*

(1) This applies mainly to city and urban areas where the services are taken underground and where the Authorities provide a connection on the cable sheath

to which the consumer's earth-continuity conductor can be connected. This is possibly the most efficient method provided that the impedance of the earth-fault path is kept sufficiently low. It applies also in rural districts, where, in addition to the phase and neutral conductors, an earthing lead is also run overhead to which the earth terminals of all consumers on the system are connected, and which at the supply end is connected to the neutral conductor and the low-voltage distribution earth electrode. The earthing conductor may also be earthed at selected points on the system and also on the consumer's premises. Provided that the conductor is of adequate cross-section, this form of protection is satisfactory, except that, should the earth-continuity conductor break and an earth-fault occurs on a consumer's premises, then possibly all consumers beyond the break from the supply may attain a dangerous potential between any exposed metalwork and the general mass of earth.

One further point concerning direct earthing is that, under fault conditions, heavy currents may flow through the earth-fault path, so that a potential may develop between the metalwork and the general mass of earth. Part of the current may be diverted through fortuitious parallel paths (that is water pipes, etc.,) and these constitute a fire risk unless the fault is rapidly cleared. Hence it is essential the protective devices are correctly related to the fault conditions. (see Regulation D.22).

(2) In many areas where the Supply Authorities have not supplied a metallic return to the medium-voltage distribution earth electrode, it is not uncommon to find direct earthing methods being used by connecting the consumer's earthing terminal to an electrode system formed in the ground adjacent to the installation. Where the soil resistivity exceeds 200 ohm-mm and soil resistivities this low are very rare, direct earthing to earth electrodes driven into the ground is normally impracticable. It is, however, quite permissible to earth to such a system, provided that the resistance is low enough to satisfy a loop-impedance test on any part of the installation.

(3) Many Authorities are now adopting Protective Multiple Earthing (P.M.E.) as a means of earthing where solid earthing is inadmissible. (Fig. 46).

In this system, the earth-continuity conductor as well as the neutral conductor at the consumer's control gear is connected to the Supply Authority's neutral conductor. Any earth current developed in a consumer's installation will, therefore, flow through the neutral conductor and back to the star point of the transformer. There is an inherent risk in that, even where no earth faults exist on a consumer's equipment, should the neutral conductor break, then all the connected metalwork on the isolated side of the break will develop a potential to the general mass of earth when load current is being carried in any consumer's installation. The system should therefore be avoided except in the case of transformers serving individual installations or from private generators in which case, the earthing system has no direct connection with other systems. The system may also be used in the special circumstances approved by the Ministry of Power or the Secretary of State for Scotland.

The danger from a broken neutral may, to some extent, be alleviated, by combining direct earthing methods with the system. Thus by the systematic earthing of every consumer's installation, and at points on the distribution system itself, including the neutral point of the secondary winding at the transformer, the resistance of the neutral conductor to earth is considerably lowered. The resistance, however, must be sufficiently low to prevent dangerous potentials being established when the prospective fault current is flowing. In the approval form granted by the Ministry of Power to the Area Boards, the overall resistance of the neutral conductor to earth must not exceed 10Ω and be as evenly distributed as is practicable.

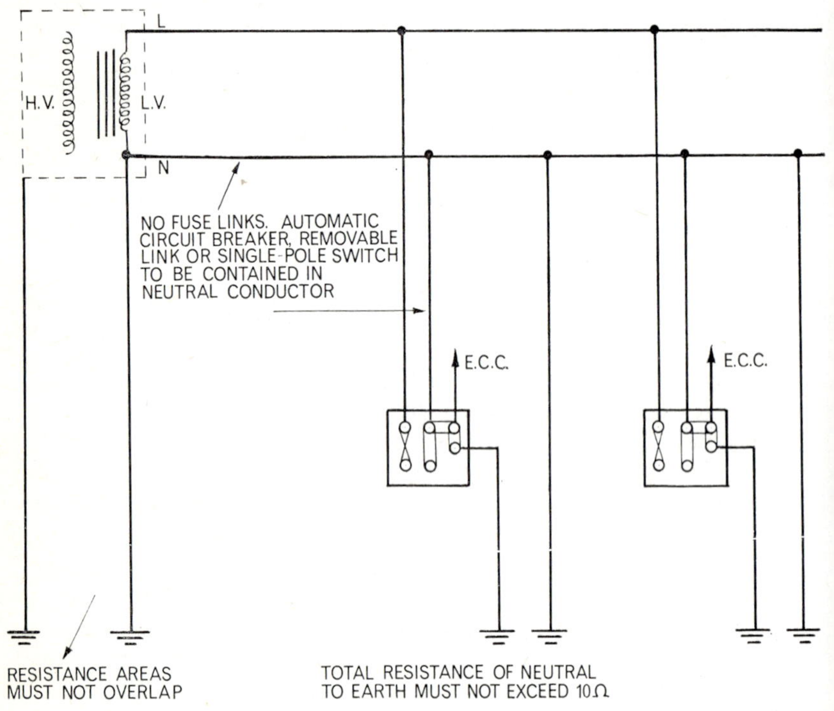

FIG. 46 *Illustrating protective multiple earthing (P.M.E.).*

A further point is that, as the neutral conductor has to carry the fault current, which might be considerable where low resistance earth faults are present, there is a danger that the neutral conductor might become overloaded, so constituting a possible fire risk. For this reason the neutral conductor must not be less in cross-section than the phase conductors.

The regulations contained in the approval form are very stringent, but they must be complied with before any P.M.E. system can be installed, one important point being that *all* the installations connected to such a system must be similarly protected.

55. *Why is it necessary to bond the consumer's earthing terminal to any gas or water service on the consumer's premises, preferably at the intake position? Why is it necessary to bond the earth-continuity conductor to certain extraneous metalwork?*

The reason for this regulation is the necessity to ensure the equality of potential between the metalwork of the various services at the intake position. Also, it is

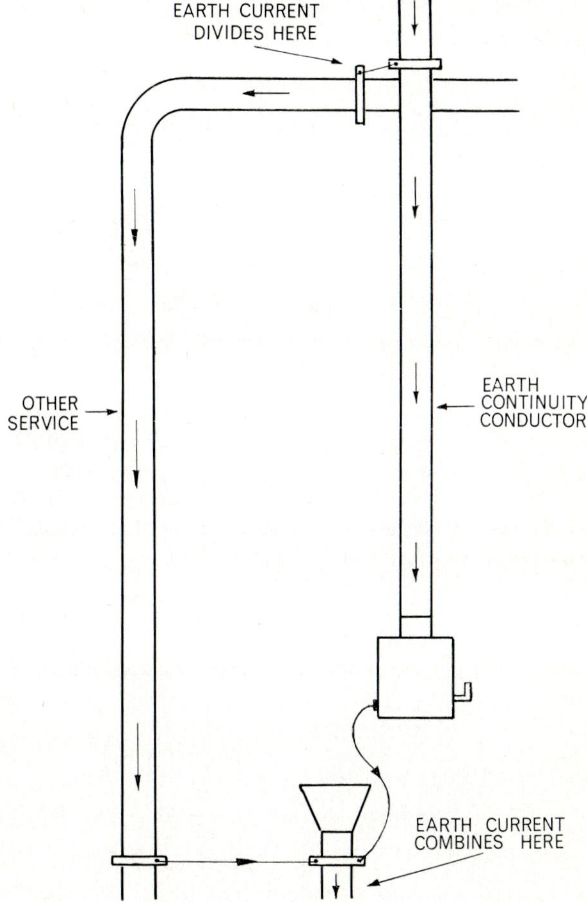

FIG 47. *Bonding the earth terminal to another service.*

essential that the metalwork of other services should not be used as an earth-continuity conductor. Bonding ensures that, should the metalwork of other services be inadvertently carrying current, the current is returned to the lead sheath of the cable at the earth electrode (Fig. 47). Similarly, the connection should be made as near as possible to the point of entry so that, under fault conditions, a relatively high potential cannot exist in the metalwork of the other service between the connection point and where it enters the ground.

The necessity for bonding the metalwork of other services to the earth-continuity conductor is because of the possibility of sparking taking place between the two metalwork systems should they inadvertently be touching when the earth-continuity conductor is carrying fault current (Fig. 48). Such sparking could puncture the sheath of a water or gas pipe. Bonding eliminates the possibility of any sparking taking place.

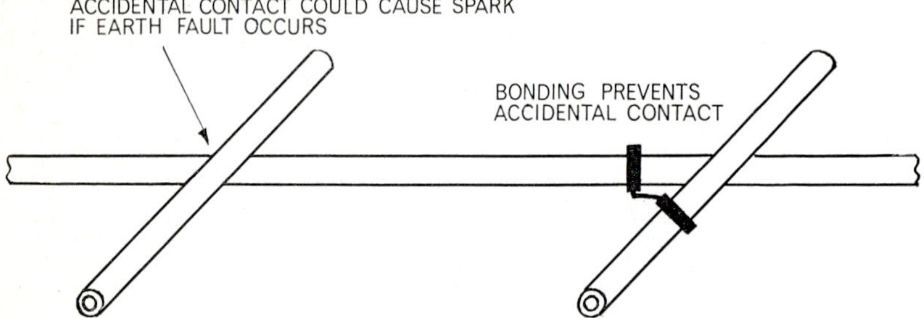

FIG. 48 *Illustrating bonding to prevent accidental contact.*

56. *State some of the main causes of high resistance in the earth-continuity conductor. Illustrate the danger of such high resistance by calculating the potential between the earthing lead and the lead sheath of the supply cable if, due to bad connection, there is an impedance of* 11 Ω *at the connection to the sheath and the remainder of the earth-fault path has an impedance of* 1 Ω. *Assume all the fault current flows through the earth path.*

Rust and corrosion; loose contacts at threads; enamel not cleaned off conduit; bad contact between plugs and socket-outlets; inefficient bonding on lead-sheathed systems; loose connections in mineral-insulated cable glands, broken flexible cords.

$$\text{Total impedance} = 11 + 1$$
$$= 12\ \Omega$$
$$\text{current} = \frac{240}{12}$$
$$= 20\ \text{A}$$

In a large majority of circuits, the fault current would not operate the protective device.

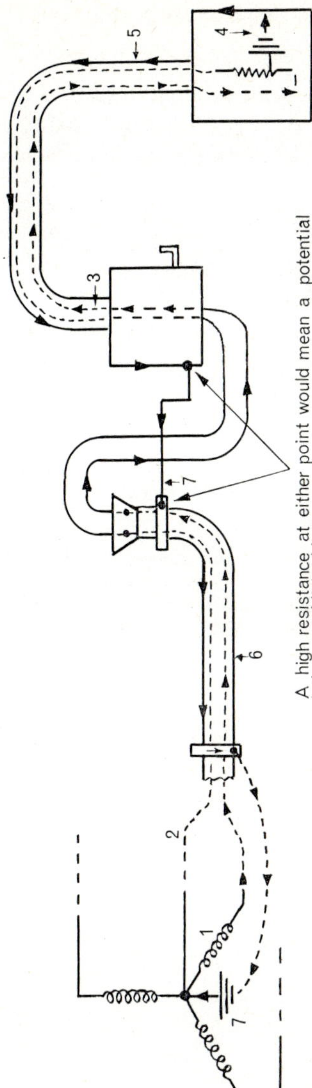

1 SUPPLY TRANSFORMER SECONDARY WINDING
2 SUPPLY AUTHORITY'S PHASE CONDUCTORS
3 CONSUMER'S PHASE CONDUCTORS
4 EARTH ON APPLIANCE
5 CONSUMER'S EARTH-CONTINUITY CONDUCTOR
6 SUPPLY AUTHORITY'S EARTH-CONTINUITY CONDUCTOR (normally lead sheath of cable)
7 EARTHING ARRANGEMENTS

Arrows denote path of earth current

FIG. 49 *Illustrating the danger of high resistance in the earth path.*

Potential between earthing lead and lead sheath $= 20 \times 11$
$= 220$ V

Thus, if the protective device was inoperative, anybody touching the earth lead and the lead sheath would receive a shock of 220 V (Fig. 49).

57. *List the various parts of the earth-fault path and tabulate the maximum values of the earth loop impedance for* 15 A, 30 A, 60 A, 100 A, 200 A, *and* 300 A *fuses,* (*a*) *with a fusing factor greater than* 1·5 *and* (*b*) *with a fusing factor less than* 1·5.

(Fig. 49).
(1) The secondary winding of the supply transformer.
(2) The Supply Authority's phase conductors.
(3) The consumer's phase conductor (mains, sub-mains and sub-circuits).
(4) The consumer's earth-continuity conductor.
(5) The earth faults.
(6) The Supply Authority's earth-continuity conductor.
(7) The earthing arrangements.

Table 2. *Relationship between fuse rating and earth loop impedance*

Current rating of fuse	*Impedance, Ω,*	
Amps	*Fuses with factor greater than* 1·5	*Fuses with factor less than* 1·5
15	5·3	6·7
30	2·7	3·3
60	1·35	1·65
100	0·8	1·0
200	0·4	0·5
300	0·27	0·33

58. *What would be the potential developed between the metal case of a* 240 V, 2000 W *boiling plate and the consumer's earth electrode if an earth fault occurs* (*a*) *between the live terminal and the case,* (*b*) *the neutral terminal and the case, and* (*c*) *at a point midway along the element and the case, if the individual parts of the earth-fault path have the following impedance values:*

Phase conductors 1 Ω.
Neutral conductors 1 Ω.
Consumer's earth-continuity conductors 1 Ω.
Authority's earth-continuity conductor 1 Ω.

Neglect the impedance of the secondary winding of the supply transformer and the earthing arrangement.

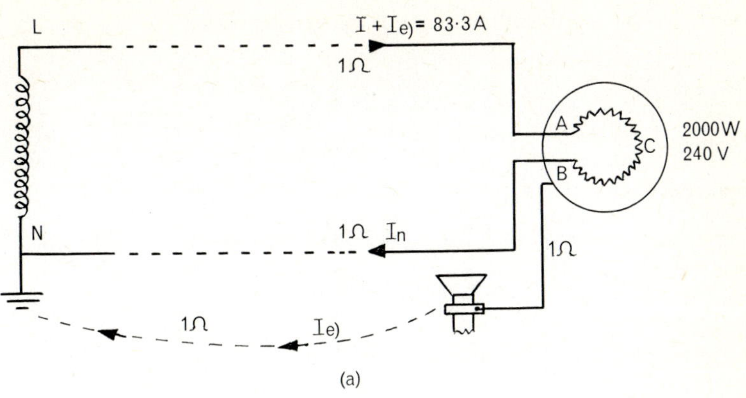

(a)

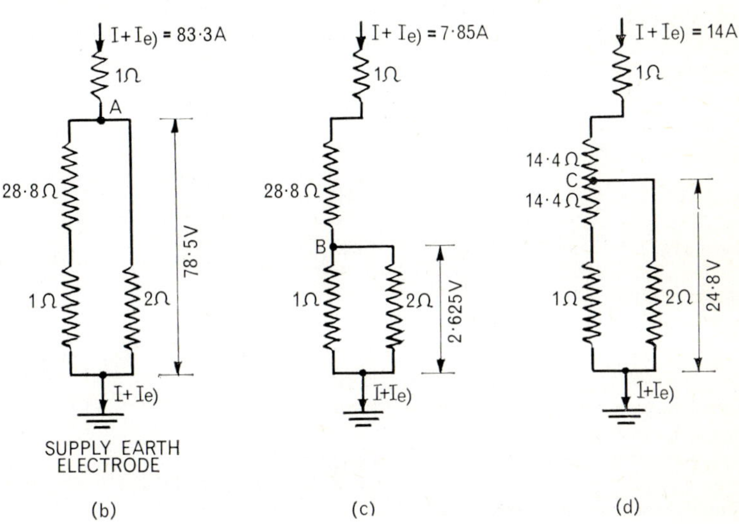

(b) (c) (d)

FIG. 50 *The circuit for question* **58.**

Fig. 50(a) illustrates the circuit arrangements and Fig. 50(b), (c), (d) the equivalent circuits.

(*a*) Resistance of element $= \dfrac{V^2}{W}$

$$= \frac{240 \times 240}{2000}$$

$$= 28{\cdot}8\ \Omega$$

Let R = the equivalent resistance of the parallel path,

then $$\frac{1}{R} = \frac{1}{29{\cdot}8} + \frac{1}{2}$$

therefore $$R = 1{\cdot}88\ \Omega$$

Total resistance $$= 1{\cdot}88 + 1$$
$$= 2{\cdot}88\ \Omega$$
$$I = \frac{240}{2{\cdot}88}$$
$$= 83{\cdot}8\ \text{A}$$

Potential between metal case and consumer's earth electrode $$= \frac{1{\cdot}88 \times 83{\cdot}8}{2}$$
$$= 78{\cdot}5\ \text{V}$$

Although quite a nasty but not normally dangerous shock might be received, the protective device should interrupt the supply almost immediately.

(*b*) $$\frac{1}{R} = \frac{1}{1} + \frac{1}{2}$$

therefore R $$= 0{\cdot}67\ \Omega$$

Total resistance $$= 1 + 28{\cdot}8 + 0{\cdot}67$$
$$= 30{\cdot}47\ \Omega$$
$$I = \frac{240}{30{\cdot}47}$$
$$= 7{\cdot}85\ \text{A}$$

Potential between metal case and consumer's earth electrode $$= \frac{7{\cdot}85 \times 0{\cdot}67}{2}$$
$$= 2{\cdot}625\ \text{V}$$

Note that 5·25 A would flow through the neutral conductor and 2·625 A through the earth-continuity conductor.

(*c*) $$I = \frac{I}{15{\cdot}4} + \frac{I}{2}$$

therefore $$R = 1{\cdot}77\ \Omega$$

Total resistance $$= 1{\cdot}77 + 15{\cdot}4$$
$$= 17{\cdot}17\ \Omega$$

$$I = \frac{240}{17{\cdot}17}$$
$$= 14\text{ A}$$

Potential between metal case and consumer's earth electrode $= \dfrac{14 \times 1{\cdot}77}{2}$

$$= 12{\cdot}4\text{ V}$$

Although, in this instance, the potential between the metal case and the consumer's earth electrode is low, it must be remembered that this potential depends upon the position of the earth fault in the appliance and the ratio of the consumer's earth-continuity impedence to the impedance of the Supply Authority's earth-continuity conductor. Thus, the nearer the earth fault is to the live terminal and the higher the ratio of the consumer's earth-continuity conductors impedance to the impedance of the Supply Authority's earth-continuity conductor, the higher is the potential between the metal case and the consumer's earth electrode. In some instances, quite a dangerous potential could be developed before either the protective device operated or the element burnt out.

59. *Describe the operation of a single-phase current-operated earth-leakage circuit-breaker and state in what circumstance it would be installed.*

The single-phase, current-operated earth-leakage circuit-breaker (Fig. 51(a)) comprises a coil-operated double-pole switch of the all-insulated type. Associated with it is a current-balanced transformer with the primary winding divided into two parts and the secondary winding (the search coil) wired in series with the operational coil of the circuit-breaker. Each coil of the main windings is connected respectively in the live and neutral conductor and are so wound that when the circuit is healthy, the currents in the two conductors are balanced and therefore, the fluxes produced by the windings are equal and opposite in direction. Hence, the resultant flux cutting the search coil is zero, the induced e.m.f. is also zero, and the operating coil of the circuit-breaker remains unenergised.

When an earth-fault occurs on the phase conductor, additional current flows through the coil associated with that conductor. Hence the flux distribution in the transformer core is no longer equal and lines of force cut the search coil, inducing an e.m.f. in it. When the earth fault is of such a value that the out-of-balance currents in the two windings generate sufficient e.m.f., the trip coil is energised sufficiently to trip the breaker.

A testing device is included which consists of a resistor and a test button wired in series, the combination being connected across the phase and neutral conductors as shown in Fig. 51(b). When the button is pressed, one of the main windings is by-passed, so producing the necessary out-of-balance current to trip the breaker. By grading the resistor, the test switch may be used to check that the device is operating at the correct order of sensitivity.

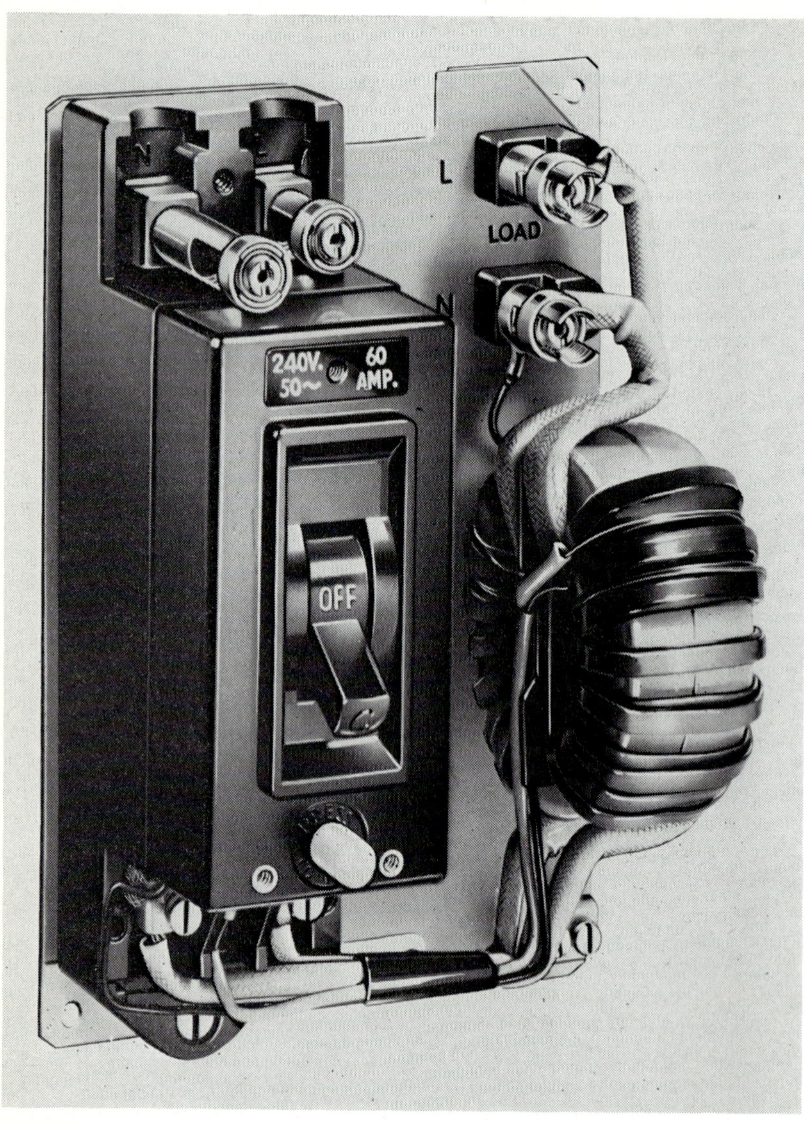

FIG. 51 (a) *The single-phase, current-operated, earth-leakage circuit-breaker.* (*Courtesy Crabtree Ltd.*)

The out-of-balance current at which the breaker will trip depends upon the rating and construction of the breaker, but typical values are 500 mA for a 30 A breaker and 1 A for a 60 A breaker, the tripping time being as low as 0·1 second.

In general, the current-operated type requires a lower loop impedance value than the voltage-operated type, that is the earth loop impedance for a 500 mA breaker must not exceed 80 Ω if the voltage of the metalwork to earth has not to exceed 40 V.

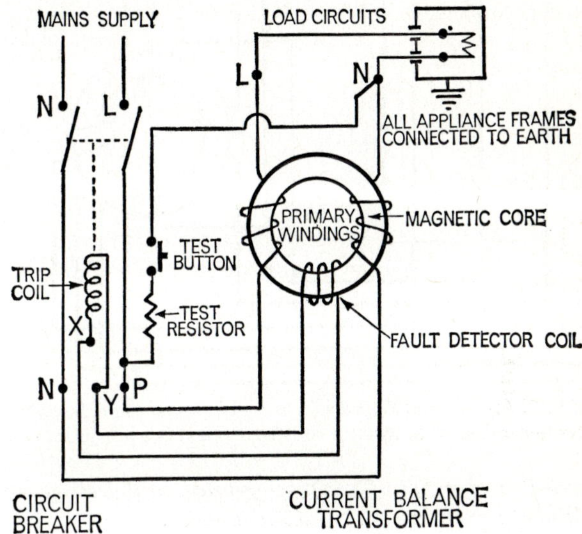

FIG. 51 (b) *The circuit of a single-phase, current-operated, earth-leakage circuit-breaker.*

Very sensitive current-operated breakers have been developed for particular applications with operating currents in the order of 1 to 2 mA, in which case, the earth loop impedance may have a value of several thousand ohms.

Current-operated protection is desirable where the installation contains several circuits each requiring discriminatory protection. This is because the operation is not affected by fortuitous parallel paths, and therefore the device protects only the circuit with which it is associated.

Also, because of its fast operating time, the circuit-breaker may be used advantageously on direct earthing systems where rapid fault clearing is essential.

The principle may also be applied to three-phase circuits; Fig. 51(c) illustrates a balanced three-phase supply incorporating earth-leakage protection. When the currents are balanced, the sum of the individual e.m.f.s is zero, and no current flows through the operating coil of the breaker. When an earth fault occurs, excess current flows in the line associated with the fault. Consequently the sum of the e.m.f.s is no longer zero, and when the fault current attains sufficient value, the current in the trip coil becomes large enough to trip the breaker.

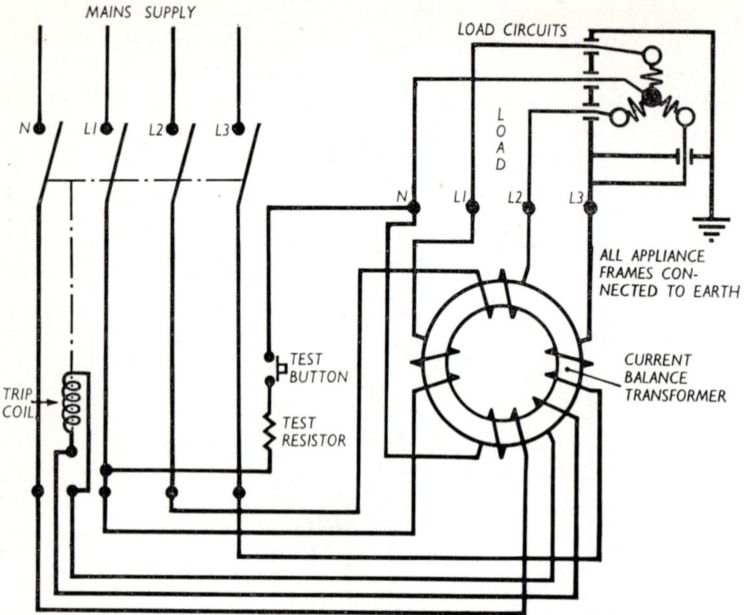

FIG. 51 (c) *Circuit diagram for three-phase and neutral supply with current-operated, earth-leakage circuit-breakers.* (*Courtesy Crabtree Ltd.*)

60. *Describe the operation of a voltage-operated, earth-leakage circuit-breaker, and state in what circumstances it would be installed.*

The voltage-operated earth-leakage circuit-breaker (Fig. 52) comprises a coil operated contactor (double-pole for single-phase and triple-pole for three-phase supplies), the coil being wired in series with the metalwork to be protected and the earth electrode. When an earth fault occurs on the installation, current flows through the coil to the earth electrode. A potential is established across the coil and the breaker trips, usually before the potential exceeds 40 V. Under these circumstances, the resistance of the earth-electrode system must not exceed 500Ω. Where the resistance does not exceed 200 Ω, the tripping voltage must not exceed 24 V.

In order to establish this potential it is essential that sufficient current is not diverted from the coil circuit, therefore, the connecting lead of the coil must be insulated, so confining any earth-leakage current to the coil circuit. Also, although there are many parallel paths to earth in a large installation, it is essential that none of these should be placed sufficiently near to the breaker earth-electrode to divert current from the coil circuit, so rendering the breaker inoperative. Although the operating current is only in the order of milliamperes, where low resistive paths exist in the vicinity of the electrode, relatively heavy currents may have to flow through the earth-continuity conductor before the breaker can trip.

The breaker also includes a test switch which, by connecting the coil through a resistor to the phase conductor and the earth-electrode, simulates earth-fault conditions.

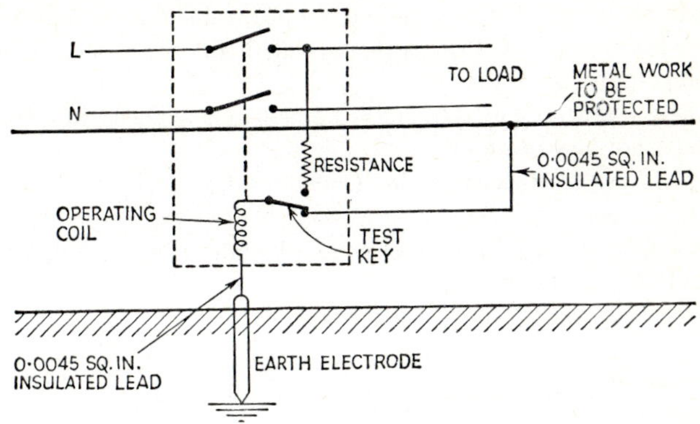

FIG. 52 *Single-phase, voltage-operated, earth-leakage circuit-breaker.*

Both types of earth-leakage circuit-breakers are used where the impedance of the earth path is so high that the normal protective devices are inoperative, but voltage-operated types can usually operate with a much higher resistance in the earth-electrode system than current-operated types. They suffer from the disadvantage of inadvertent tripping which may be caused by:

(1) Transient earth faults, such as where cookers, etc., have been standing idle, so that when they are initially switched on the earth current is sufficient to trip the breaker.

(2) Dormant neutral earth faults where, should an earth fault exist, and the voltage in the neutral rises to about 20 to 30 V, this is sufficient to trip the breaker.

(3) Faults imported from other parts of the installation, through parallel paths.

61. *Describe how portable appliances may be suitably protected against dangerous earth-leakage currents.*

Where the appliance is operating at mains voltage and any metalwork associated with it is exposed, it is essential that this metalwork is effectively earthed. Therefore particular attention should be paid to the following:

(1) Correct connections are essential at both plug and socket, that is, no switch must be connected in the neutral conductor.

(2) Good contact is essential between the earth-contact tube in the socket-outlet and the earth pin of the plug.

(3) Effective cord grips must be provided to prevent excessive pull on the cord and the green wire should preferably be a little longer than the other conductors.

(4) Insulated separators must be provided in plugs to separate the cores.

(5) Where extension leads are used, correct polarity must be ensured.

(6) Three-phase socket-outlets should not be installed in the same installation as single-phase socket-outlets if plugs associated with the latter can be plugged into the three-phase socket-outlets.

(7) Damaged flexible cords should be immediately replaced and twisted taped joints should not be permitted.

(8) Appliances, together with the flexible cords, should be periodically examined and tested.

(9) Radio-interference suppressors should be built into the plug or appliance and not in the flexible cord.

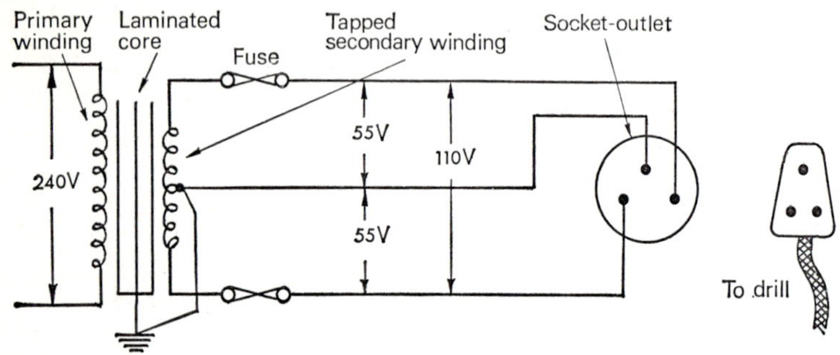

FIG. 53 *Reduced voltage supplies for portable appliances and handlamps.*

Safer methods of operating portable appliances are to either use double-insulated appliances, in which case earthing is unnecessary, or to operate them on reduced voltage supplies (Fig. 53). The preferred voltage is 110 V, with possibly a tapping at 25 V for supplying portable handlamps. If the centre-point of the

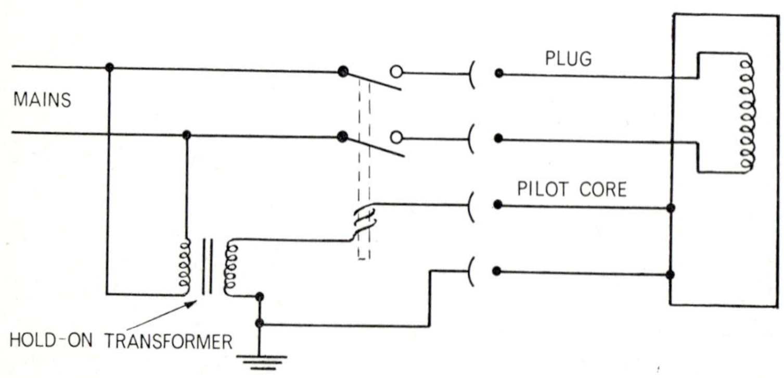

FIG. 54 *Monitored earth-leakage protection of a portable appliance.*

transformer is tapped, a maximum voltage of 55 V to earth is ensured and, if necessary, additional protection may be provided by connecting the earth lead through the trip coil of an earth leakage circuit-breaker. This is essential in boiler-houses and similar situations.

Where it is essential that any break in the earth-continuity conductor be quickly detected, the earth-continuity conductor in the cord may be monitored through a trip coil (Fig. 54) which disconnects the circuit should the earth-continuity conductor be broken.

62. *Describe the methods which may be used for obtaining an earth electrode system in the ground, and state how the resistance of the system may be reduced.*

Plate electrodes. This type of electrode is usually employed where the resistivity of the top soil is relatively low and possibly where it is kept damp. The plates are generally of copper, not less than 2·5 mm thick, or of cast iron, not less than 12·5 mm thick, and, in practice, it is found that they need not exceed 1·5 m by 1·5 m in area. The plates are normally set vertically at such a depth that they are surrounded by damp soil.

To lower the resistance, plates may be connected in parallel, but care must be taken to ensure that the resistance areas of each plate do not overlap. To be effective, the spacing, therefore, should be at least 10 m, but in practice, it is found a spacing of 2 m does not increase the combined resistance by more than 20 per cent. Where cast-iron plates are used, the joint between the plate and any copper earthing lead must be protected from corrosion by dissimilar metal action by coating it with a good bitumastic paint.

Rod or pipe electrodes. Usually plates are uneconomic to install and would be mainly used only where the current loading is liable to be high. For most locations either rods or pipes would be used. In practice, it is found that the most effective system is when the dimension in one direction is large in comparison to the other two dimensions, and therefore, as the ratio of the length to the cross-sectional area is high, increasing the diameter of the rod has little effect on the resistance. The resistance of the soil however, does diminish rapidly for the first 3 m, then less as the rods are driven deeper.

The resistance of the system may be reduced by using rods or pipes in parallel, and, for a spacing of 3 m, the resistance of two rods in parallel is little more than half that for individual rods.

Rods may be of 12·5 mm or 20 mm diameter copper and in lengths from 1·5 m to 5 m. For deeper driving, the rods are screwed, so enabling them to be coupled together. Where the resistivity of the top soil is low, far better results are obtained by paralleling rods rather than by deep driving into the soil, but where the resistivity is high and it is possible to reach soil of relatively low resistivity then deep driving may be advantageously employed.

Strip electrodes. This type of electrode can be advantageously used where high resistivity soils lie below shallow surface soils of low resistivity, or where rocky sub-strata lie below the top soil. The electrode may be of untinned copper strip not less than 25 mm by 1·5 mm cross-section, or it may be of bare copper as used for overhead lines. The strip must be buried at least 500 mm and even deeper where there is a possibility of mechanical damage from agricultural implements. The resistance may be reduced by laying strips in parallel, in a spiral, or radial formation from a point. Where parallel strips are used, the resistance of two lines spaced 3 m apart is normally found to be less than 65 per cent of the individual resistance of either strip.

Chemical treatment of soil. Another method of reducing the resistance of the electrode resistance is by chemically treating the soil. One of the most commonly used chemicals is salt. Where rod electrodes are installed, a hemispherical hole is excavated round the top of the electrode. Approximately one hundredweight of damp salt is then poured into the hole and covered with top soil. Rainfall assists the salt to percolate into the top soil, so reducing its resistivity in the area round the electrode. For strip electrodes, the top soil is mixed with salt before replacement. The effect, however, is only temporary, lasting approximately 18 months to 2 years, depending largely upon the drainage of the soil. Also, the corrosive action of the soil is increased by salting, resulting in a more rapid deterioration of the electrode.

Another method sometimes used for either plate or strip electrodes is to surround the electrode with coke breeze. The method is not recommended, however, as the corrosive effects are increased, not only to the electrode, but to any cable sheath, water pipe etc., to which it is bonded. In such cases it is better to use rod electrodes. Also, copper electrodes must not be used, owing to the corrosive nature of the sulphur contained within the coke breeze, and where a copper lead is connected to a cast-iron plate, the connection must be suitably protected by a coat of bitumastic paint. In addition, coke treated electrodes must not be installed within 6 m of other metalwork unless the latter is suitably protected against corrosion.

Other methods of reducing the resistance are by surrounding the electrode by powdered charcoal instead of coke, or by treating the soil with solutions of calcium chloride, sodium carbonate, or by using sulphates, such as copper sulphate.

7

Testing

Polarity—earth-continuity—earth-loop impedance—insulation—earth-leakage circuit-breakers—earth electrode—ring mains—faults in metal-sheathed cables.

63. *Why is it essential that the polarity of circuits is correct and how may it be verified?*

The following examples illustrate how danger may occur through the polarity being incorrect.

(1) A switch in the neutral conductor instead of the live conductor of a lighting circuit means that irrespective of the position of the switch contacts, there is always full potential between one conductor and any adjacent earthed metal.

(2) The incorrect polarity of either a plug, or socket with which it is associated, may mean that an appliance such as an electric fire may appear to be switched off, yet full potential may be present between the element and adjacent earthed metalwork.

(3) Where Edison-type screw lampholders are being connected to the supply, it is essential to ensure that the phase conductor is taken to the centre contact. Should it be taken to the screwed part of the lampholder, it is possible for an operative removing the lamp to touch the metal part of the lamp cap and receive a dangerous shock. It should be remembered that even a slight shock to anyone working on the top of a ladder may prove fatal.

Where the installation is connected to the supply, testing of the polarity of conductors is relatively simple. To test the polarity of a socket-outlet a neon tester may be used. Possibly the simplest method is to connect a test lamp between the L and E terminals of a suitable plug, which then may be used to test each socket-outlet in turn (Fig. 55).

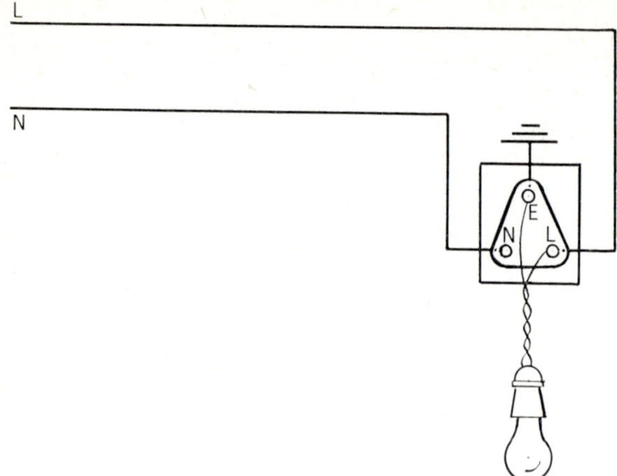

FIG. 55 *Testing the polarity of socket-outlets.*

To test the polarity of conductors at switch positions, a neon tester or test lamp may be used. All lamps must be removed, and, with the switch cover removed, each terminal is tested, polarity being correct if the neon tester lights. If a test lamp is used, a return connection to either the neutral or the earth-continuity conductor must be provided. If the lamps are difficult to remove, the polarity may be tested by opening and closing the switch. If the polarity is correct, the test lamp will remain alight irrespective of the switch position. If the polarity is incorrect, the lamp will light with the switch open but will be extinguished when the switch is closed.

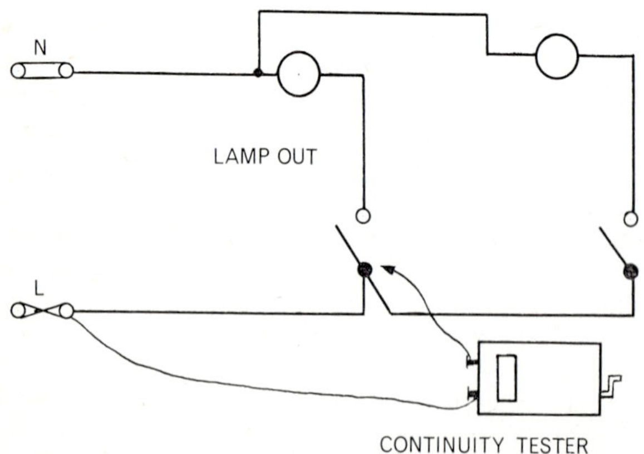

FIG. 56 *Testing the polarity of switches.*

Where the installation has not been connected to the supply, the polarity of conductors may be tested by using a bell and battery, an ohmmeter, or a continuity tester. One test lead is connected to the phase conductor at the intake position, then, to test socket-outlets, the other test-lead of the instrument is connected to the L terminal of a suitable plug, which is then connected to each socket-outlet in turn. The polarity of conductors at switches is tested by removing the lamps and connecting the instrument between the terminal of the switch and the phase conductor at the intake position (Fig. 56). In each case, a zero reading indicates that the polarity is correct.

64. *Describe how the effectiveness of the earth-continuity conductor may be tested* (1) *using the loop-impedance tester*, (2) *using the d.c. hand tester, and state what precautions are necessary.*

(1) The test of an earth-continuity conductor should be made with alternating current at the frequency of the supply and of magnitude approaching 1·5 times the rating of the final sub-circuit under test, except that the current need not exceed 25 A. For this test, the installation must be connected to the supply. Where possible, one end of the earth-continuity conductor is connected to the loop-impedance tester, the other end being connected through a cable of known resistance to the other terminal of the tester, which is set for continuity testing. A voltage, which must not exceed is 40 V, is then applied to the circuit. It is normally satisfactory if, subject to the requirements for the total loop impedance, the voltage/current ratio does not exceed 1 Ω.

(2) Where it is not convenient to use alternating current to test an a.c. installation (such as where the supply is not connected), direct current may be applied provided that it has been ensured that no inductor is incorporated in the circuit. Where a d.c. hand tester is used, one end of the earth-continuity conductor is connected to one terminal of the tester, and the other end is connected to a cable of known resistance. The opposite end of the cable is then connected to the other terminal of the tester. It is normally satisfactory where steel conduit or pipe forms part or the whole of the earth-continuity conductor if the resistance does not exceeding 0·5 Ω. Where the earth-continuity conductor is composed entirely of copper, copper-alloy or aluminium, the resistance should not exceed 1 Ω.

When measuring the resistance of the auxiliary cable with the loop-impedance tester, it is essential to unroll it otherwise an abnormally high impedance (not resistance) is obtained because of the reactive effect of the coil.

Although the test is based on the probability of a relatively high test current, this value does depend on the impedance of the circuit. If the impedance of the earth path is high, for an applied voltage of 40 V, the current will be low.

When testing from an a.c. supply, the effect on the impedance for a high current is more pronounced, hence the reason why a higher value of impedance is permissible with this test.

Where a d.c. hand-tester is used, the current is normally in the order of 10 to 100 mA, but as the test current is being rapidly reversed, the effect of any inductor in the circuit would be to give false readings.

If imperfect joints are suspected in the earth-continuity conductor, the resistance varies with the test current. It is recommended, therefore, that a d.c. resistance test should be made, first with a low current of approximately 200 mA, then a second test at a current of not less than 10 A, then a third test at low current again. The open-circuit voltage (possibly from a battery) should not be less than 30 V and any substantial variation in the readings obtained will indicate faulty joints.

The earth-continuity conductor may also be tested by two other means given in Appendix 6 of the I.E.E. Regulations, but whichever test is applied it is important to remember that any figure quoted in the tests is subject to the loop-impedance test being satisfactory. Referring to Table 2, page 104 it can be seen that where a 300 A fuse is used to protect the circuit, the impedance of the whole earth path must not exceed 0·33 Ω, which is below the lowest value for the earth-continuity conductor in any of the tests specified in Appendix 6.

65. *Describe the object of earth-loop impedance testing and state how it is carried out using the neutral-earth loop tester. What errors may be introduced and how may they be minimised?*

The regulation concerning the earth-loop impedance is the most important one in the Regulations, as it concerns the safety of life against shock, and property against the possibility of fire. Should the earth-fault path be defective, the fuse or excess current circuit-breaker may fail to operate and disconnect the faulty circuit when an earth fault develops. The test, therefore, is to prove whether the impedance of the earth-fault path is low enough to operate the protective devices, whether these consist of fuses or excess-current circuit-breakers.

Fig. 57 illustrates the wiring diagram for a commercial type neutral-earth loop-impedance tester and Fig. 58 shows simply how it can be used to test the earth-loop path to a socket-outlet.

The tester contains three double-pole change-over switches, the purpose of which are:

Switch 1 is to test the polarity of the circuit. This is important as should the polarity be incorrect, almost the full supply potential could be impressed across the earth-continuity conductor.

Switch 2 is used to minimise the effect of current flowing in the neutral from other sources than that of the test circuit, possibly from the consumer's own installations and also from other consumer's neutral conductors.

Switch 3 is used to test the loop-impedance value of the earth path, and also the impedance of the earth-continuity conductor, a socket-outlet also being incorporated for the testing of the effectiveness of the earthing of appliances.

FIG. 57(a) *Neutral-earthed loop impedance tester and* (b) *the wiring diagram for the loop impedance-tested.* (*Courtesy Meritus* (*Barnet*) *Ltd.*)

Assuming a 13 A socket-outlet is being tested, the tester is connected to the socket-outlet, the polarity is verified by means of switch 1, and switch 3 is switched to loop-impedance testing. Switch 2 is turned to its first position, the press-button is operated and the reading is noted. Switch 2 is then changed to its second position, the process being repeated and the reading noted. By this means the voltage drop in the neutral is added to the applied voltage when the

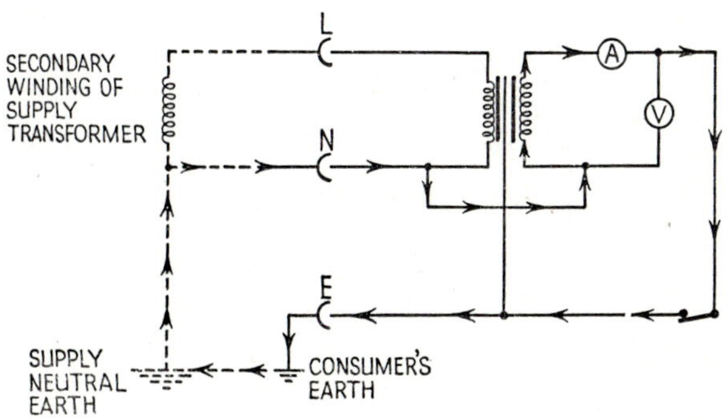

FIG. 58 *Carrying out a neutral-earth loop-impedance test.*

A C MAINS
N
L
E
SW1
1
LN
OK
2
LN
REV
RL
PB
RELAY
CONT.
GL
T1
M
SW2
1
2
SW3
1
2
F1
SK
TX
EC

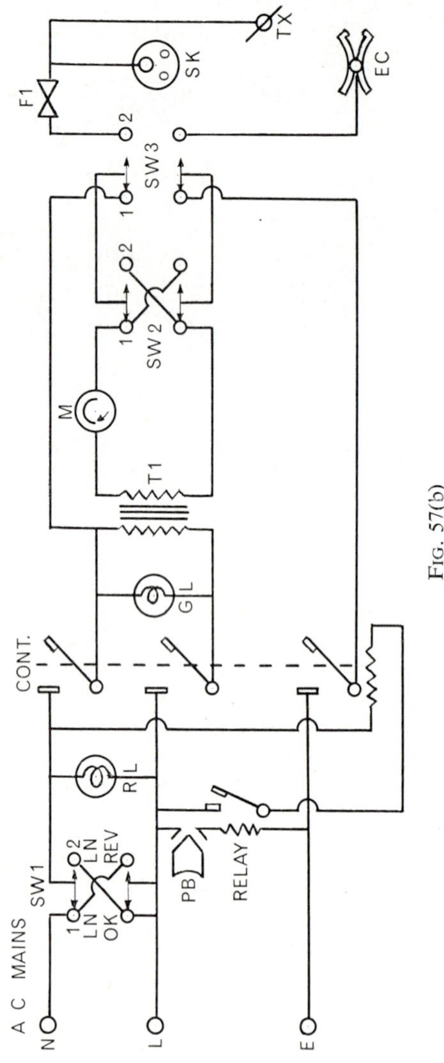

FIG. 57(b)

switch is in one direction and subtracted in the other direction, so virtually eliminating the effect of current flowing in the neutral conductors from its own and other sources.

By finding the mean of these two values, and dividing it into the *supply* voltage, the actual current is found which would flow should a short-circuit occur between the phase conductor and the earth-continuity conductor at the socket-outlet. The obtained value must exceed the value required by Regulation D.22 of the I.E.E. Regulations (see question **54**).

Although switch 2 minimises the effect of voltage drop in the neutral conductors, should there be an appreciable delay before the second reading is taken the current may have altered, resulting in a slight inaccuracy of the mean value. A more serious error could occur if dormant earth faults exist on the distribution system other than the one under test. These faults would provide parallel paths to the testing circuit and may virtually by-pass part of the testing circuit.

Misleading results may also occur where reduced neutrals have been installed, but as the neutral would have a correspondingly higher resistance than the phase conductor, this is not detrimental to the test. Also, where the impedance of the testing transformer is relatively high compared to that of the circuit under test, allowances must be made for the impedance of the windings.

Any cross-bonds to other services must be made before the test is carried out. This is important in that the current is high and fires have been caused where this precaution has been neglected, Although cross-bonds at the intake position reduce this hazard, they do not altogether eliminate it. Where it is suspected that inadvertent contact is made between two services (possibly beneath floorboards), before the cross-bond is made, a test should be undertaken, using a hand-operated tester, to ensure that the two services are not touching.

66. *Describe how a loop-impedance test may be carried out using a line-earth loop-impedance tester and state what precautions must be taken in carrying out the test.*

Many electrical contractors now prefer to use this method of testing since, as the earth current flows through the phase conductor, it is thought better to use the same conductor for testing purposes. As, however, the test current is relativele high, the period during which it flows has to be limited, in the order of milliseconds. Also, on P.M.E. systems, since the neutral is frequently earthed, the neutral-earth type is unsuitable, and the line-earth tester would therefore by preferred.

Whichever type of line-earth tester is used, the fundamental principle is the same in that a resistance, usually in the order of 1 Ω, is connected in series with the earth-fault path. An artificial fault is simulated and the potential developed across the resistor is applied across some form of indicating device. Fig. 59 illustrates a simplified diagram of this type of testing.

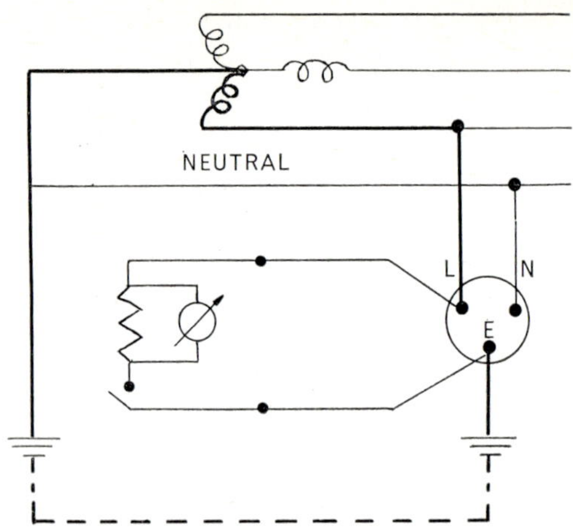

FIG. 59 *Simplified diagram of line-earth testing.*

FIG. 60 *Line-earth loop-impedance tester, direct reading type. (Courtesy Evershed & Vignoles Ltd.)*

There are many types of line-earth testers, but in the one illustrated in Fig. 60, a resistor of approximately 10 Ω is connected in the earth-fault path. This enables a current of approximately 10 A to flow for a limited period of time (about 0·2 seconds), the duration being controlled by a time-delay relay.

When the operating switch is pressed, current flows through the phase conductor, through the resistor, and so through the earth path. The potential established across the resistor is applied to a ballistic galvanometer. As this potential depends upon the value of the earth-loop impedance, the instrument may be calibrated in terms of the impedance of the earth path. A voltmeter is normally included which may be used as a polarity indicator and also to read the voltage of the supply.

To carry out the test, the tester is connected to the socket-outlet to be tested, the polarity switch is operated to ensure that the polarity is correct, then, as the needle is not spring controlled, a re-set button is pressed which sets the needle to the zero position. This operation is essential for every test, otherwise inaccurate results may be obtained.

The operating switch is pressed so enabling the impedance of the earth-fault path to be found.

As with the neutral-earth tester, by dividing this value into the supply voltage, the fault current may be found.

Apart from ensuring that the polarity is correct, it is necessary to ensure that the earth path is not so defective as to develop dangerous potentials in the earth-fault path. Hence, the effectiveness of the earth-continuity conductor should be tested first.

Although the type illustrated in Fig. 60 gives a direct reading in ohms, other forms of line-earth testers merely give an indication whether the circuit is satisfactory or not.

67. *Describe the purpose of insulation testing and state how these tests are carried out before an installation is connected to the supply.*

The purpose of the insulation is to confine the flow of electricity to the conductors and to prohibit any leakage of current which might cause danger to life or property. Insulation testing, therefore, is to prove if the insulation round the conductors is satisfactory with regard to leakage from the conductors to adjacent metalwork, and also between phase conductors, and between the phase and neutral conductors.

One important supply regulation concerns itself with this point and specifies that the Supply Undertakings shall not permanently connect an installation to the supply unless they are satisfied that the connection would not cause a leakage exceeding one ten-thousandth part of the maximum current of the installation.

The requirements of the I.E.E. Regulations, however, are normally more stringent, and the reading for any of the prescribed insulation tests must not be less than 1 MΩ. Thus, where an installation has a maximum current of 60 A, in accordance with the Supply Regulations, the leakage current may be up to 6 mA, whereas the requirements of the I.E.E. Regulations stipulate that the leakage current must not exceed 0·24 mA on a 240 V supply. Hence, if the requirements of the I.E.E. Regulations are satisfied then the requirements of the Supply Regulations will also be satisfied.

It should also be realised that although a new installation with an insulation reading of 1 MΩ would be connected to the supply, this would be considered to be a poor reading as, when the installation was retested, possibly twelve months later, the reading would have fallen below 1 MΩ due to the presence of damp, dust and dirt.

Because of reactive effects, it is necessary to use d.c. testers for insulation testing, and the Regulations specify that the testing voltage should be twice the working voltage, but need not exceed 500 V for medium voltage supplies. The 240 V insulation tester, therefore, cannot be guaranteed to give satisfactory results on 240 V supplies unless the reading obtained is far above the permissible value.

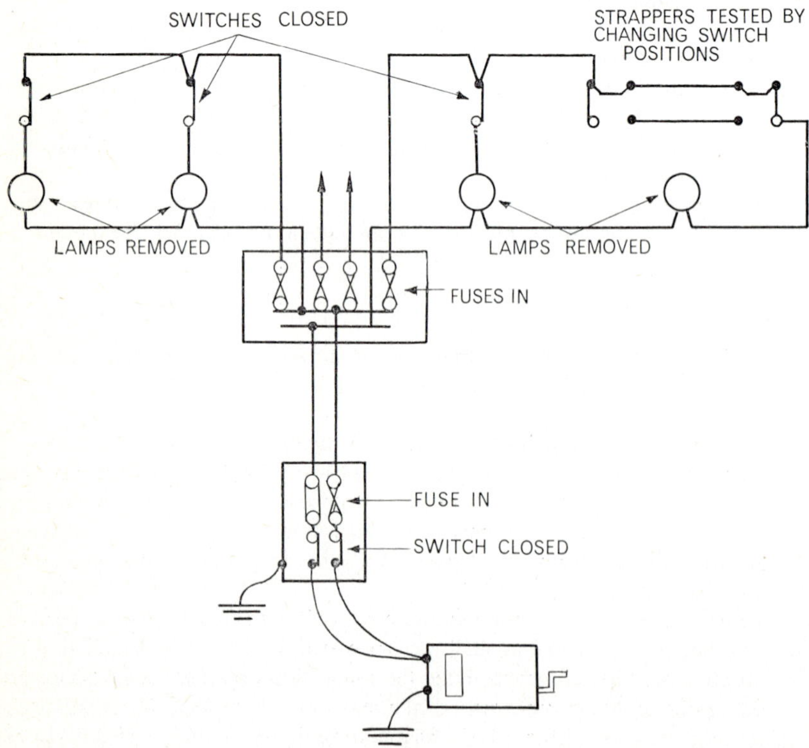

FIG. 61 *Insulation testing.*

One insulation test (Fig. 61) consists of putting in all fuse links, closing all switches, including the main switch, and connecting all poles or phases of the wiring, except on earth concentric systems, to one terminal of the insulation tester, the other terminal of the tester being connected to the earth electrode. The reading must be not less than 1 MΩ.

If the reading is less than 1MΩ, and the installation contains less than 50 outlets, then the faulty section must be located and rectified, but if the installation has considerably more than 50 outlets, it may be divided into sections each containing not less than 50 outlets, and each section should be tested separately. The reading for each section should not be less than 1 MΩ.

The necessity for dividing a large installation is that, where a low reading is obtained, it ensures that the low reading is not confined to one small part of the installation, possibly even to the one circuit. Only division of a large installation can prove this.

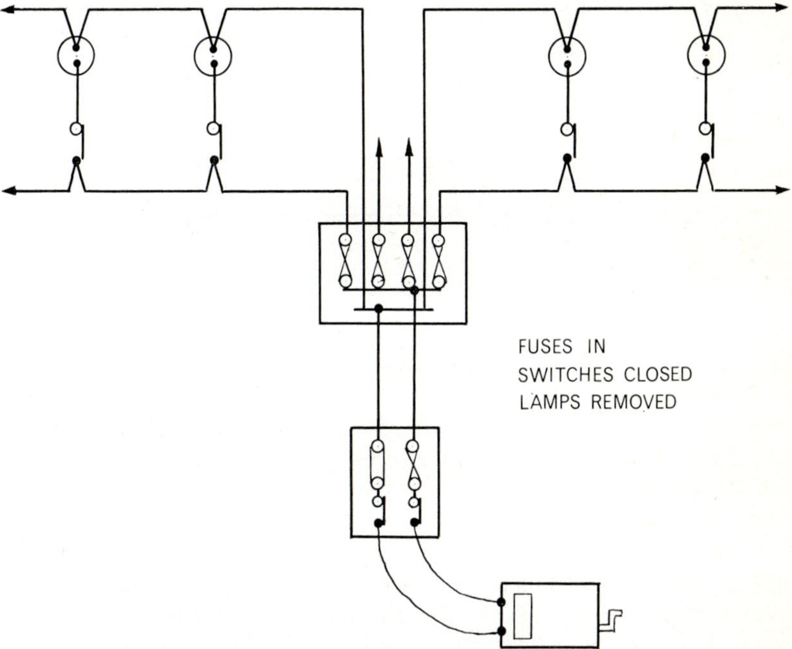

FIG. 62 *Insulation test between conductors.*

A second method (Fig. 62) is to test the insulation between all phase and neutral conductors by connecting all the conductors connected to any one pole or phase of the supply to one terminal and, in turn, to the other test terminal all the conductors connected to the other poles or phases of the supply. The insulation reading again should not be less than 1 MΩ.

All lamps must be removed and all current-using apparatus disconnected, and all local switches controlling lamps or apparatus must be closed. Where removing lamps or disconnecting apparatus is impracticable, it is permissible to test the installation with them in position but with the switches open.

Also, by leaving the lamps in, the continuity of the circuit may be tested by closing the switches while the tester is still connected.

68. *How is the insulation resistance of an installation affected by* (1) *the voltage grading of the cables,* (2) *the number of outlets,* (3) *the lengths of run, and* (4) *the connection of fixed appliances?*

(1) When an insulated cable is lying against earthed metalwork, such as conduit, wherever the cable touches the metalwork, there is a small leakage current, normally too small to be measured, passing from the cable through the insulation to the metalwork (Fig. 63). These leakage paths combine to form an equivalent resistance, the value of which decides the value of the insulation resistance of the installation, and which, in turn, is dependant upon the voltage grading of the cable.

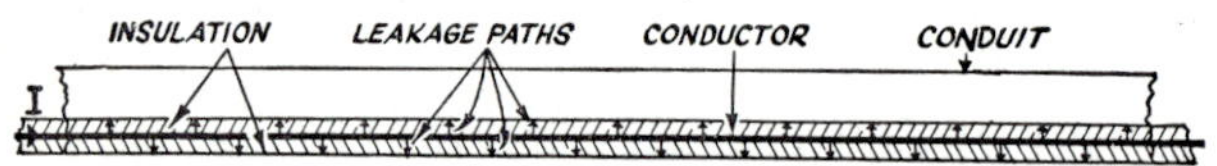

FIG. 63 *Illustrating possible leakage paths between a cable and conduit.*

For normal installation work p.v.c.-insulated cables are obtainable now in the one category only of insulation, the 250/440 V grade and the 660/1100 V grade. As the thickness of the insulation the grading depends on, the insulation resistance also increases for a given length of cable, and one would expect to obtain a higher value for the insulation test, where 660/1100 grade cables have been used than where cables with a thinner insulation have been installed.

(2) Although not strictly correct, it can normally be assumed that the more outlets connected on an installation, the greater is the number of leakage parallel paths through the insulation. Hence, the permissible value for the insulation resistance is based on the number of connected outlets. An outlet consists of every point, and every switch, except where the switch is included in a socket-outlet, appliance or lighting fitting, where it is counted as one outlet.

(3) The longer the length of run, the greater is the number of parallel paths through the insulation to earthed metalwork, hence as the length of run increases, the insulation of the cable becomes lower. It is possible to install so much cable on an installation that it is virtually impossible to obtain a satisfactory reading for the entire installation.

(4) Many appliances, such as boiling-plates, electric cookers, etc., where the heating element is very close to the metalwork, have leakage currents which, although not dangerous, have the effect, especially where the appliance has been standing idle, possibly for a long time in a damp situation, of reducing the insulation resistance of the installation to a low value, well below the permissible value. Hence, the appliance may be disconnected from the installation and be tested separately, in which case the value must be not less than 0·5 MΩ.

69. *Describe the tests that should be made to test the effectiveness of a current-operated earth-leakage circuit-breaker.*

The operation of a breaker depends only on the difference between the currents in the live and neutral conductors, and is therefore not affected by any parallel paths in the earth system.

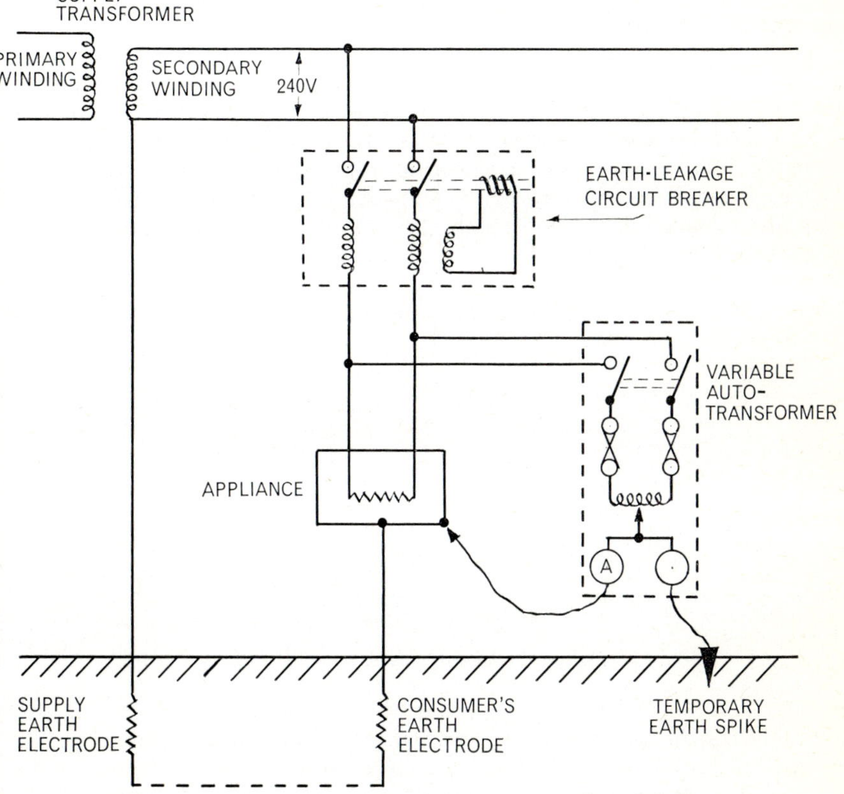

FIG. 64 *Testing the effectiveness of a current-operated, earth-leakage circuit-breaker.*

Where an integral test switch and graded resistor are provided, they may be used to test the operation of the breaker. To find the out-of-balance currents milliammeters may be connected in the live and neutral conductors.

Where an integral testing switch is not included, a test (Fig. 64) should be made similar to one for the voltage-operated breaker. The preliminary precautions are the same, but in this instance, the output from the transformer is gradually increased and the circuit-breaker should trip before the current reaches 110 per cent of the nominal setting of the circuit-breaker. Should the current required to trip the breaker exceed this, the cause should be investigated and rectified before retesting.

In no circumstances must the voltage between the protected metalwork and the general mass of earth exceed 40 V, either during the test or during normal operation. Where this occurs, the resistance of the consumer's earth electrode system must be reduced. If this is found to be impracticable, a voltage-operated, earth-leakage circuit-breaker would have to be installed.

70. *Describe how the resistance of an earth-electrode system may be measured.*

To measure the resistance of an earth-electrode system, the secondary winding of a double-wound transformer, with an output not exceeding 40 V, is connected to the electrode to be tested X and an auxiliary electrode Y placed well outside the resistance area of X (Fig. 65). A third electrode Z is placed mid-way between X and Y. Both Y and Z should be of 12·5 mm mild-steel diameter bar and driven about 1 m into the ground. An ammeter is also connected in series with the secondary winding while a voltmeter is connected between X and Z. The resistance of the earth-electrode system is the voltage drop measured on the voltmeter divided

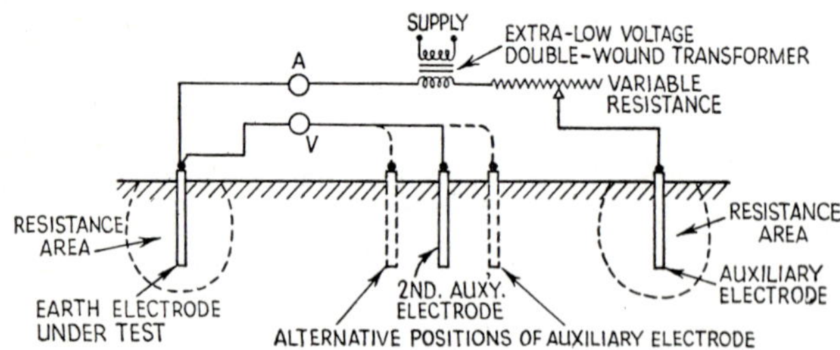

FIG. 65 *Measuring the resistance of an earth-electrode system.*

by the current measured on the ammeter. To ensure that the reading is a true one, the electrode Z is moved first, 6 m nearer to X then 6 m farther away from X, two further readings being taken. If the three readings substantially agree, the mean value may be taken as the value of the earth-electrode system. Should there be considerable variation, however, the distance between X and Y must be increased until a satisfactory result is obtained. For a single-electrode system, a satisfactory distance between X and Y is about 30 m. Where the resistance of X is low in comparison with that of Y, the resistance area of the auxiliary electrode is increased, so that a larger separation is necessary.

At power frequencies, the resistance of the voltmeter must be high compared to that of the auxiliary electrode, that is should the resistance of Z be about 500 Ω, then the voltmeter resistance should be at least 10 000 Ω for an accuracy of ±5 per cent.

This method is not altogether satisfactory if there are any stray currents in the ground at the same frequency, and, in such cases, it is preferable to use a hand-driven generator. This type generates direct current and includes a rotary current reverser and a synchronous rectifier mounted on the shaft. Thus alternating current may be applied to the test circuit (d.c. must not be used on account of electroyltic action in the ground) and the resulting potentials rectified and applied to the measuring instrument which is normally calibrated in units of resistance. Stray currents will cause the needle to waver, but the wavering will disappear if the speed is increased.

71. *Why is it necessary to test the continuity of ring-circuits, and how may such tests be carried out?*

A disadvantage of the ring-circuit of socket-outlets is that should a break occur in the ring, two radial circuits are formed to replace the ring circuit. This could lead to the overloading of one radial circuit. Thus, in Fig. 66 if a break occurs at X, the full load is carried by the radial circuit A, whereas no current flows through the radial circuit B. Periodic testing is necessary, therefore, to ensure the continuity of the ring.

If the circuit is connected to the supply, the simplest method is to remove the fuse, disconnect one line, one neutral and one earth-continuity conductor, then to replace the fuse, and test with a voltmeter between the disconnected phase conductor and the disconnected neutral conductor and then between the phase conductor and the disconnected earth-continuity conductor. Full voltage readings will indicate that the continuity of conductors is satisfactory. Alternatively a test lamp may be used.

If the supply has not been connected, then all the conductors of the ring circuit at the distribution board must be removed. Both ends of the phase conductors, the neutral conductors, and the earth-continuity conductors are connected in turn to the terminals of either a bell-and-battery set, or a continuity tester.

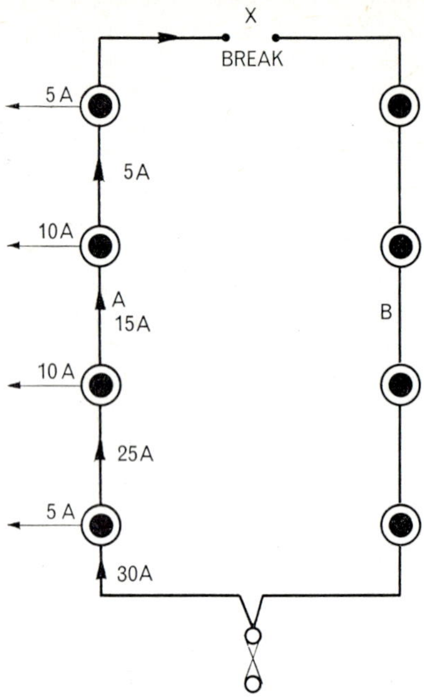

FIG. 66 *Illustrating the effect of a break in a ring main circuit.*

72. *Describe the tests that should be carried out on an appliance such as an electric kettle to prove its effectiveness.*

Many appliances are connected to the supply by means of a three-cored cable and plug, and should the appliance or the wiring system be defective, then it is possible for anyone handling the appliance to receive a dangerous shock.

It is essential therefore, that both the appliance and the wiring system should be periodically tested. To ensure that the appliance is in satisfactory working order, the following tests may be made with an insulation tester (Fig. 67).

(1) To test whether the insulation resistance between current-carrying conductors and the earth-continuity conductor is satisfactory, the latter is connected to one terminal of the tester, and the other two conductors of the flexible cord are connected to the other tester terminal. A reading well in excess of 1 MΩ should be obtained. If the resistance is low, the appliance connector should be removed to find if the low reading lies in the flexible cord or the appliance.

(2) To ensure that there is no low resistance between current-carrying conductors of the flexible cord, with the appliance connector removed, each conductor is tested in turn (L and N; L and E; and N and E). Again the reading

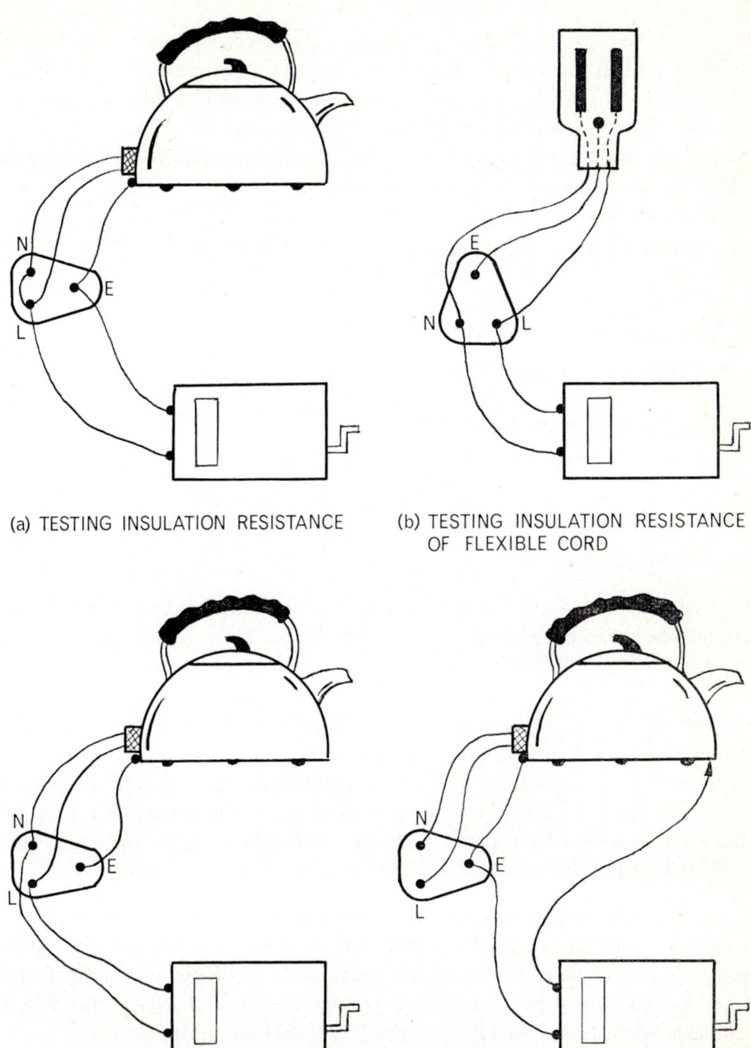

FIG. 67 *Testing an electric kettle.*

should be well in excess of 1 MΩ. If there is a low reading, further tests should be made, first with the appliance connector disconnected, then with the plug disconnected from the flexible cord.

(3) To ensure continuity of the circuit, the appliance connector is replaced in the appliance, then the current-carrying conductors are connected to the respective terminals of a tester, now set for continuity testing. If the continuity is sound, the reading should be zero. If this is not obtained, the appliance connector should be removed and the lead and appliance tested separately to find whether the fault lies in the appliance or the flexible cord.

Fixed appliances may be tested similarly, but testing becomes more laborious as the appliance is not so easily disconnected from the supply.

73. *Describe the types of fault which occur on wiring systems and state how they may be located.*

The faults may be classified as follows:

(1) low insulation resistance between any conductor and earth;

(2) low insulation resistance between conductors;

(3) resistance in the circuit causing dimming of lamps;

(4) cross connections in the wiring system.

(1) Assuming all lamps have been removed and all switches opened, each circuit is disconnected and tested individually until the faulty one is found. Next, the insulation of the live conductor to earth is tested with the insulation tester. If the instrument registers a low reading with the switches open, the fault must lie on one of the switch feeds. These are divided in turn and tested separately until the faulty section is found.

If the reading is satisfactory with the switches open, then each of the switches must be operated respectively until the faulty section is found. If the neutral conductor has a low resistance with all the lamps removed, it too must be divided, possibly first at the half-way point, until the faulty section is located.

(2) If, with the lamps removed and the switches opened, a low reading is obtained between the phase and neutral conductors, sections must be disconnected progressively until the faulty section is found. If the reading is satisfactory with the switches open but unsatisfactory when they are all closed the switches should be opened progressively until the faulty section is found.

(3) If the entire lighting on one final distribution board is dimmed, the fault lies between the boards and possibly a section board or the main board. Starting at the intake position and testing progressively, a voltmeter will quickly indicate where the fault lies, possibly at a faulty contact in a distribution board. If, however, only one circuit is faulty and all the lights on that circuit are dimmed, the fault lies between the first point and the board. Testing with the voltmeter

will detect this, the fault again possibly being caused by a faulty contact at the fuse. If the contact is satisfactory, a full voltmeter reading should be obtained between the fuse and the neutral link, and therefore the voltage between the neutrals and switch feed at the first outlet should be tested. A low reading will indicate that either the phase or neutral conductor is faulty, and testing separately will quickly discover the faulty cable. If only part of the lights on the one circuit are dim, this can be checked by removing the fuse and finding which other lights are extinguished. Testing with a voltmeter at both the first of the faulty lights and both sides of its associated switch should locate the fault.

(4) Where colour-coded cables are used, these faults should not occur, but occasionally do through carelessness. Where m.i.m.s. cables are left unidentified, the chance of a cross-connection is much more likely.

If there is a cross-connection between the phase and neutral conductors, the fuse will rupture. On a three-plate ceiling rose system, this would probably be caused by a cross-connection at one of ceiling roses. Disconnecting the conductors at the board and testing with an insulation tester will locate the fault by progressively disconnecting the ceiling roses.

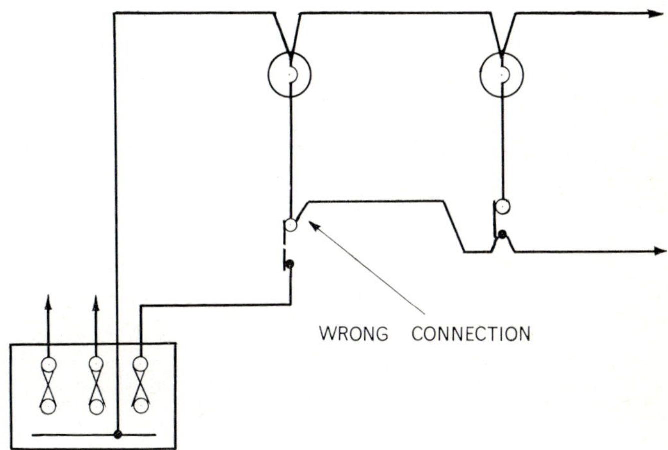

FIG. 68 *Incorrect connection of switch feeds.*

The same fault can occur on a loop-in system, but it is easier to find as the opening of the ascsoiated switch should disconnect the faulty section.

In some instances a light fails to light when switched on, but lights when a switch associated with another light is switched on. This is because the switch feed serving the faulty circuit has been connected to the switch wire of the other switch, instead of the switch feed (Fig. 68).

Another type of fault is where two or more lamps are connected in series when the switch is closed. This is due to a neutral conductor at the first point of the faulty section being connected to a switch wire instead of the other neutral conductor (Fig. 69). Where two lamps of the same rating are connected in series, they will both light with a reduced but equal brilliancy, but if the ratings vary, the light outputs of each will also vary. If one lamp is connected in series with several more lamps in parallel, then the lone lamp will light and the others will be virtually extinguished. Removing the lamps progressively will result in a diminution of the light output of the single lamp with a corresponding increase in the light output of the remaining lamps in the parallel circuit. The cross-connection, in this case, would be found to be at the single lamp.

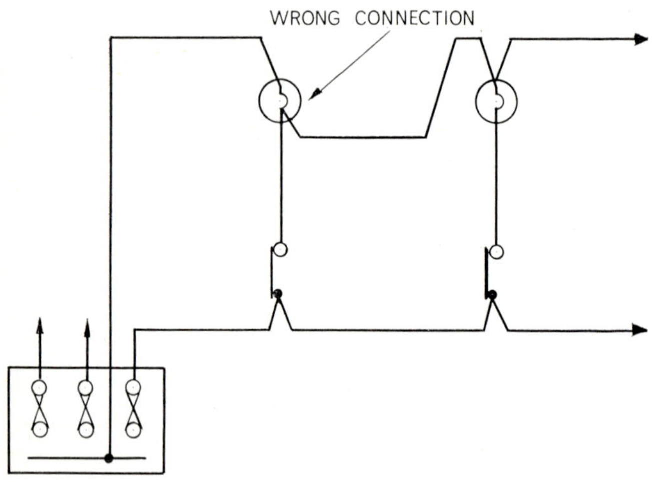

FIG. 69 *Incorrect connection of neutral conductors.*

74. *Using a Wheatstone bridge how may a fault be located* (1) *on a three-core metal-sheathed cable with a low resistance to earth on one core, and* (2) *a twin metal-sheathed cable with a low resistance on both cores to earth?*

The faulty cable is first disconnected from the remainder of the system at both ends. Insulation and continuity tests are then carried out at each end, each core being tested separately to earth and also to the other cores for insulation resistance. If there is one sound core, this may be used for testing the continuity of the other cores, but if all the cores are faulty, an auxiliary cable must be used. When all the cores have been tested for insulation and continuity, a diagram should be made illustrating the exact nature of the fault.

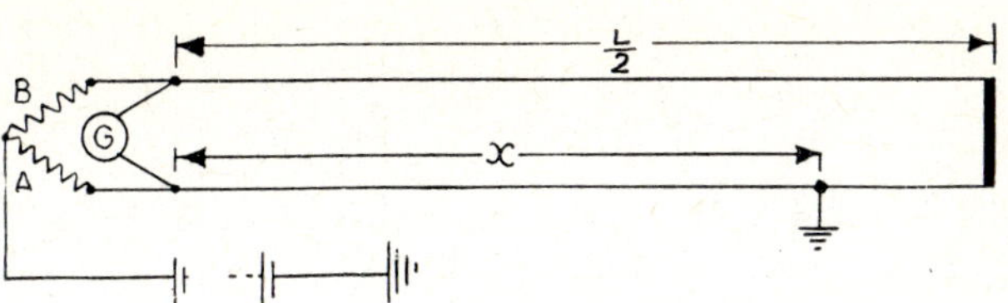

FIG. 70 *Testing for a fault on a cable using a Wheatstone bridge.*

To carry out the test, referred to as the Murray-loop test, the Wheatstone bridge, galvanometer and battery are connected to the faulty core and one of the sound cores, the two cores being solidly connected together at the end remote from the instruments (Fig. 70). Only two of the resistances of the bridge are required as the other two resistances of the bridge are formed from the faulty cable. The accumulator (6 V is ample for a low resistance fault) is connected between the mid-point of the bridge and earth and, as the cable is down to earth, this connection may be made to the lead sheath. The variable resistance is adjusted until the bridge is balanced (that is, the galvanometer reading is zero) and, if x is the distance of the fault from the instrument connection as shown in Fig. 70, then it may be found from the formula:

$$x = \frac{L \times A}{B + A}$$

where L is the full cable loop (the length of both cores), A is the value of the variable resistor and B the value of the fixed resistor. Note that as the cable is of uniform cross-section and resistivity, the formula may be stated using units of length for the cable instead of units of resistance.

Where all cores of the cable have a low resistance fault to earth, then as the cores of the cable are by-passed by the fault, an auxiliary cable must be used as the sound core. Hence the connections are as shown in Fig. 71. The method is the same as that described previously, but unless the auxiliary cable is of the same cross-section as the faulty cable the equivalent length of the auxiliary cable must be found in terms of the faulty cable.

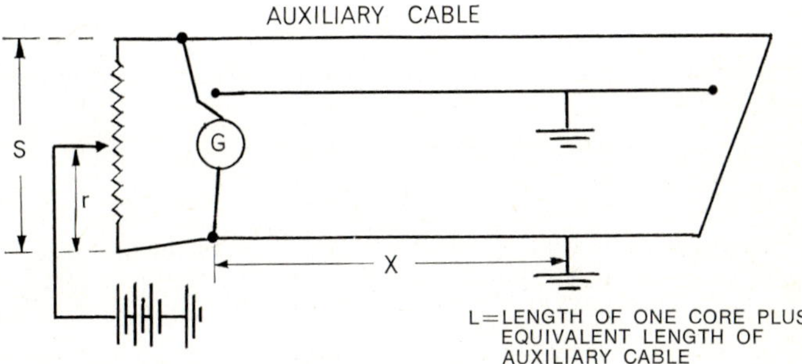

FIG. 71 *Testing for a fault using an auxiliary cable.*

If the faulty cable has a cross-section of 240 mm², and a 400 m long 120 mm² cable is used as an auxiliary cable, the length in terms of the faulty cable, (that is, which would have the same resistance if a 240 mm² cable had been used) would be 200 m.

The bridge is balanced by varying the resistance until the galvanometer reading is zero and x is found from the formula:

$$x = \frac{L \times r}{S}$$

where L is the length of one core plus the equivalent length of the auxiliary cable.

75. *How may a low resistance fault between two cores of a three-core, metal-sheathed cable be located using a slide-wire bridge?*

The method of finding short-circuit faults is similar to that used for locating earth faults, but as there is no connection to the lead sheath the instruments are connected to the lines as shown in Fig. 72, the cable loop being formed by the

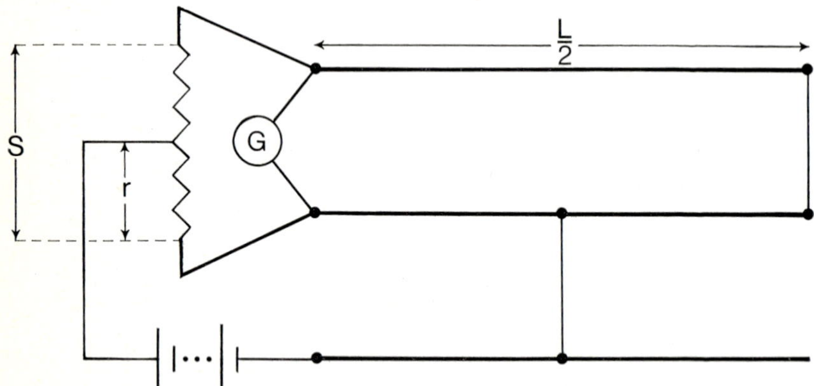

FIG. 72 *Testing for a fault on a cable using a slide-wire bridge.*

sound core and one section of each of the faulty cores. In this instance, a slide wire-bridge is used in place of the Wheatstone bridge mainly because of its simplicity of use.

The sound core is connected to one of the faulty cores at one end, then when the instruments have been connected at the opposite end, the bridge is balanced by adjusting the movable contact on the slide wire until the galvanometer reading is zero. The following equation is then true:

$$x = \frac{L \times r}{S}$$

where S is the full length of the slide-wire bridge, r is the distance from the point on the bridge where balance occurs to the faulty core, and L is the cable loop which now consists of the sound core and a length equivalent to the length of one of the faulty cores.

If there is a fault on all cores, then again an auxiliary cable would have to be used.

If there is any doubt about the position of the fault, then the test should be repeated from the other end of the cable.

76. *How may an open-circuit be located on a twin core cable, using (a) a galvanometer and (b) a capacitance bridge?*

In this test, the faulty cable is charged by connecting the cable via a switch and battery to earth and charged to about 100 V for 15 seconds. The core is then discharged through the galvanometer to earth, the deflection being noted and referred to as C_1. Where the other core is sound, it may be charged and discharged similarly, the deflection of the galvanometer again being noted and referred to as C_2. As the galvanometer deflections are proportional to the capacitance, the position of the fault may be located in terms of the two deflections. If x is the distance to the break in the cable (Fig. 73) and L is the length of one core, then:

$$x = \frac{C_1 L}{C_2}$$

Where no sound core is available, the readings may be taken from both ends of the faulty cable, in which case C_2 now becomes the reading from the second end. The expression becomes:

$$x = \frac{C_1 L}{C_1 + C_2}$$

If one sound core is available, a check on the first reading should be made by taking readings from the opposite end of the faulty core.

Where the insulation has a low resistance, possibly because of the presence of moisture, it is preferable to use an a.c. capacitance bridge for testing purposes

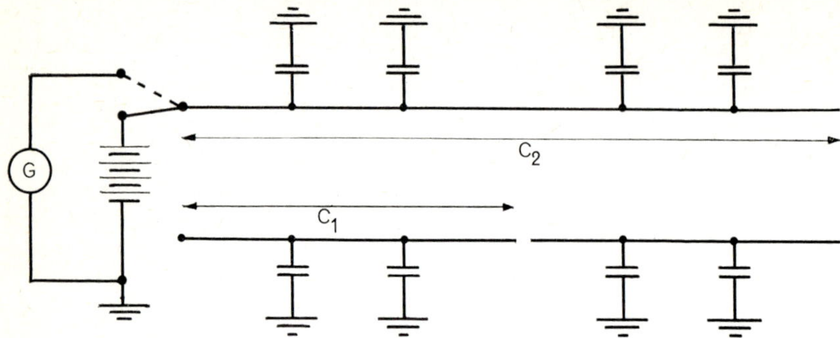

FIG. 73 *Locating an open-circuit fault on a cable using a galvanometer.*

(Fig. 74). To provide an audio-frequency supply an oscillator is connected across two arms of the bridge, the third arm being formed by a variable resistor and variable capacitor connected in parallel, while the fourth arm is formed by the capacitance of the cable itself.

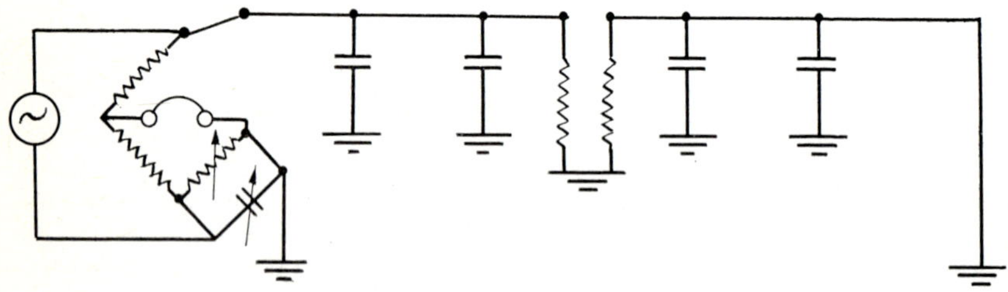

FIG. 74 *Locating an open-circuit fault on a cable with a capacitance bridge.*

A phone may be connected across the bridge and the bridge is balanced by alternately adjusting the variable resistor and capacitor until no sound is received on the telephone receiver. The capacitance of the cable is then found from the reading of the variable capacitor.

During the test, any core not being tested should be earthed, and any broken core should be earthed at the end remote from the testing position.

77. *How may high resistance faults be located between cores and between cores and earth in buried cables?*

This method of testing, referred to as induction testing, is normally used for detecting core-to-core faults or core-to-earth faults where the impedance is relatively high, but preferably not more than 100 Ω. Where the impedance exceeds this value, it is necessary to reduce it by breaking it down with a high a.c. or d.c. voltage, or by passing heavy alternating current through the fault from the low-voltage mains through a limiting resistor. For low-voltage supplies, the second method is to be preferred.

The procedure is to connect an oscillator, energised possibly from a 12 V battery, to the faulty cores, so enabling a supply at approximately 1000 Hz to be introduced into the faulty cable. An external field is thereby produced, the maximum intensity of which varies from point to point owing to the helical lay of the conductors. A search coil is used which, as it moves along the cable route, picks up the signal and transfers it to an amplifier before passing to the headphones. Because of the variation in the magnetic field strength, the intensity of the signal also rises and falls periodically as the search coil follows the cable route. Where non-metallic cables are installed, the signal disappears as the search coil passes over the fault, but the test is not successful where metal-sheathed cables are installed as current flows in each direction at the fault position. The method is particularly useful, however, for tracing the cable run of a heating system of buried metallic-sheathed cables, the actual position of the fault being located by means of the capacitance bridge.

8

Transformers

Single-phase transformers—auto-transformers—instrument transformers —three-phase transformers—testing.

78. *Describe the full operation of a single-phase transformer off and on load and state how the power factor of the load affects its performance.*

When the transformer is connected to the supply with the secondary winding open-circuited, the current in the primary winding produces a magnetic flux in the iron core which in turn induces an e.m.f. in the primary winding which is almost equal to the applied voltage but opposite in direction so the current in the primary winding is kept to a low value. The alternating flux also induces in the secondary winding an alternating e.m.f. of the same frequency as that of the supply.

When an external circuit is connected across the secondary winding, current flows through it, reducing the effective value of the flux in the iron core. This in turn reduces the impedance of the primary circuit allowing an increase of current which restores the flux to a value only slightly less than its no-load value.

The effect of power factor is to reduce the useful load that can be carried by the transformer. The rating of a transformer is governed by the temperature rise of the windings, which, in turn, is dependent, not on the useful current, but on the total current, which is inversely proportional to the power factor.

Also, as the copper losses are proportional to the square of the total current, the efficiency is slightly reduced as the power factor of the output falls, the equation for the efficiency becoming:

$$\text{percentage efficiency} = \frac{\text{output kVA} \times \cos\phi}{(\text{output kVA} \times \cos\phi) + \text{losses}} \times 100$$

79. *Describe the construction and operation of an auto-transformer and state its advantages and disadvantages.*

The auto-transformer (Fig. 75) is a transformer with a single winding, part of which is common to both primary and secondary circuits. In the double-wound transformer, power is transferred from one winding to the other by electromagnetic induction, but in the auto-transformer, some of the power is transferred directly. Thus, in Fig. 75, the current in the common part is the difference between

SINGLE-PHASE TRANSFORMERS AND AUTO-TRANSFORMERS

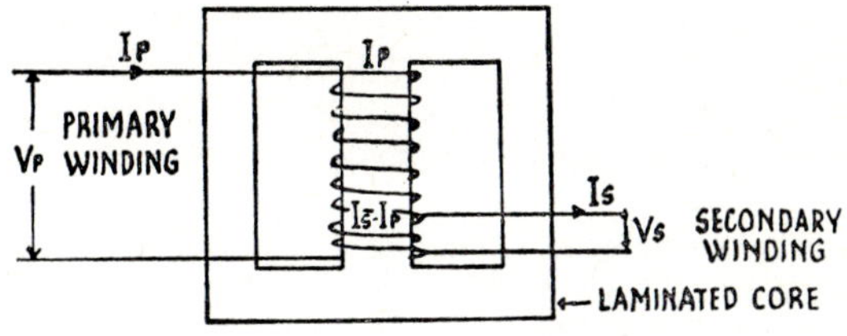

FIG. 75 *The auto-transformer.*

the secondary and primary currents. The ratio of the secondary voltage to the primary voltage again corresponds with the transformation ratio of the transformer, that is the ratio of the number of turns on the primary winding to those on the secondary winding.

The advantage of the auto-transformer is that, for similar outputs to those of the double-wound transformers, there is a considerable saving of copper in the windings, especially where the transformation ratio approaches one-to-one, where the saving is a maximum.

One disadvantage of the auto-transformer is that, in certain installations, such as bell or signalling circuits, if it was wrongly connected (Fig. 76) there would be full mains voltage between one of the conductors at the bell push and the general mass of earth. Another disadvantage is that any disturbance in the system connected with one winding is more likely to affect the circuits connected to the other windings than it would if the transformer had two separate windings

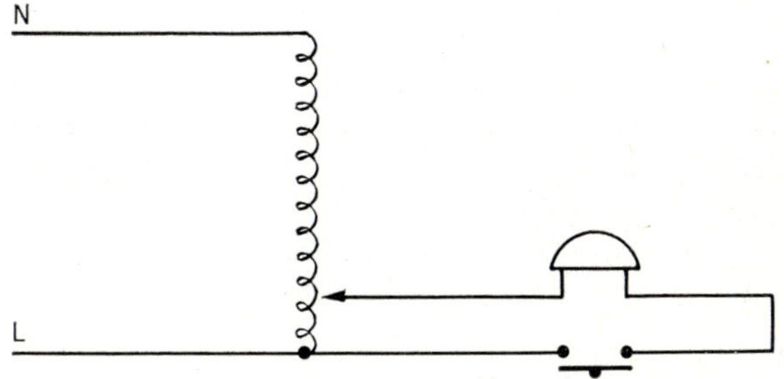

FIG. 76 *Illustrating the danger of incorrect connection of an auto-transformer.*

80. *Describe the difference between a step-down transformer and a step-up transformer.*

The transformation ratio of a transformer is given by the equation

$$\frac{V_p}{V_s} = \frac{N_p}{N_s} = \frac{I_s}{I_p}$$

Cross-multiplying

$$V_s = \frac{V_p \times N_s}{N_p}$$

$$I_s = \frac{N_p \times I_p}{N_s}$$

If the voltage has to be stepped up, then the ratio of N_s to N_p must be greater than unity. Thus, the primary winding of a step-up transformer has relatively few turns compared to the secondary winding. For an input of constant power the output current must decrease as the output voltage is increased. Thus for high voltage/low current outputs small diameter wire can be used.

It is impossible to operate most electrical equipment at the high voltage used in transmission lines, and, therefore, step-down transformers are employed to reduce the voltage to a safe working value. In this case N_s to N_p must have a value less than unity. Thus the primary winding has a large number of turns of relatively low cross-section, while the secondary winding has a low number of turns of relatively high cross-section.

81. *What is meant by the terms Percentage Regulation and Percentage Voltage Reactance and how does the reactance of the transformer affect the breaking capacity of switchgear under fault conditions?*

Like other electrical apparatus, the transformer does not maintain a constant secondary output voltage as the load varies, the voltage tending to fall as the load increases. This behaviour is referred to as the regulation of the transformer. A transformer is said to have a good regulation value when there is only a small variation of secondary voltage with load.

The regulation is defined in terms of the difference between the secondary voltage on no-load and full-load at 75° C and at constant primary voltage. It is usually expressed as percentage of the voltage of the secondary winding on no-load, and is known as percentage regulation. It is defined as the difference between the secondary voltage on no-load and on full-load, divided by the voltage on no-load, and multiplied by 100. Hence,

$$\text{Percentage regulation} = \frac{E - V}{E} \times 100$$

Where E is the no-load value and V the full-load value.

FIG. 77 *Voltage instrument transformer.* (*Courtesy Foster Transformers Ltd.*)

In a transformer, the reactance voltage is the voltage at the rated load in quadrature with the current. It is usually expressed as a percentage of the rated secondary voltage. For most transformers, this is in the order of 5 per cent.

The reactance of the windings of a transformer is high compared with the resistance, the ratio being in the order of 10 to 1. Therefore, when a fault, such as a short-circuit, occurs on the system, the reactance serves as a current limiting device in the circuit, so preventing the fault current and kVA from rising to excessively high values.

82. *Describe the applications of instrument transformers in electrical circuits. Describe how the current transformer differs from the conventional transformer and state what precautions must be taken when it is in use.*

Instrument transformers may be classed in two groups, (1) voltage transformers, and (2) current transformers.

The voltage transformer (Fig. 77) is used to transform high voltages to voltages more suitable for operating the required instrument, and also, where necessary, to segregate the instrument entirely from the supply. The construction of the voltage transformer is similar to that of the power transformer. Where the primary voltage is high, the transformer dimensions may be quite large although, owing to the difficulty of insulating the primary and secondary windings, it may have only a relatively low output.

Current transformers are used where the current in the circuit to be measured is too high to pass safely through the instrument. Thus, although a circuit may be carrying possibly 100 A by means of the current transformer, only a low current (normally 5 A for full-scale deflection) is flowing through the instrument.

Apart from being used for measuring circuits, instrument transformers are used for the protection of distribution systems and installations against excess current and dangerous earth-leakage faults.

The construction of the current transformer (Fig. 78) differs from the voltage transformer in that, as the transformer is used to transform currents, the primary winding is connected in series with the load to be measured, and therefore the impedance must be low enough not to affect the normal output of the load. The primary winding, therefore, may consist of few turns of relatively large cross-section. For circuits of exceptionally high current ratings, it may comprise a straight through bar or conductor.

Alternatively, the primary may consist of the cable carrying the load current being passed through the centre of the core and secondary winding. The insulation of the cable must be capable of withstanding any strain imposed upon it and, as a further precaution, it may be shrouded by a porcelain bushing.

In current transformers, errors are produced by the primary current supplying the magnetizing current and the core losses, and therefore the design of the core should be such as to keep the magnetizing and loss currents as low as possible.

The secondary windings of a current transformer should never be left open-circuited while the primary circuit is carrying current as this would result in an increase in the flux density which would increase the iron losses sufficiently to cause excessive heating in the core and windings. A dangerously high e.m.f. might also be generated in the secondary winding. As the primary winding is in series with the load, the primary current is dependent upon the load current and not the secondary current. Hence, excessive current does not flow through the secondary winding even when it is short-circuited.

FIG. 78 *Current instrument transformer.* (*Courtesy Foster Transformers Ltd.*)

83. *Describe how the windings of three-phase transformers may be connected, and explain, giving reasons, which system is used for low- and medium-voltage distribution.*

The windings of three-phase transformers may be connected in several ways. Fig. 79 illustrates the windings and how they may be connected, together with the correct symbols and phasor diagrams.

From Fig. 79, we can see that three-phase transformers are divided into four groups, Group 3 and Group 4 being the most widely adopted for transforming the voltage down to that required for the three-phase, four-wire standard distribution system.

The reason for this is that many services such as lighting supplies require a connection to the neutral conductor. Both the transformers shown in Phasor Groups 31 and 33 and Phasor Groups 41 and 43, fulfil this requirement, allowing

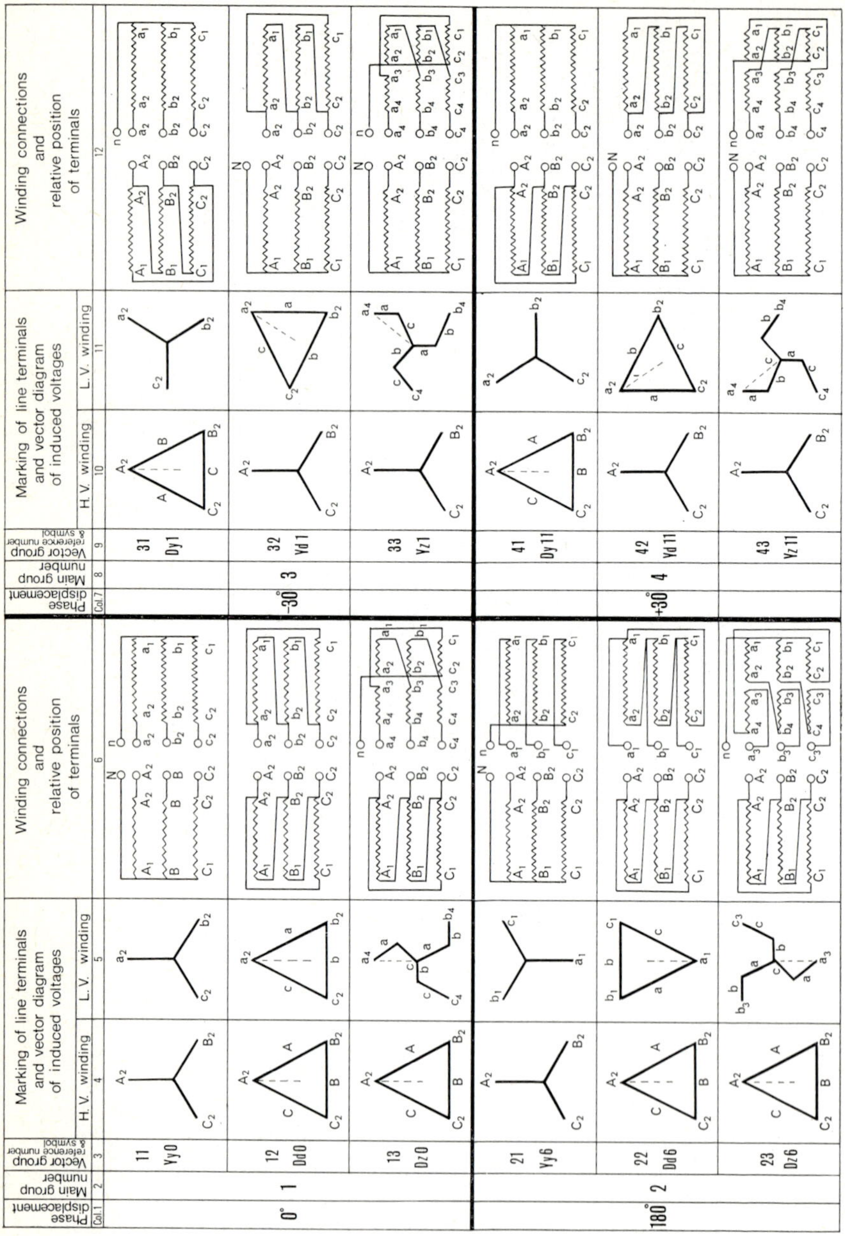

Phase displacement (Col. 1)	Main group number (2)	Vector group reference number & symbol (3)	Marking of line terminals and vector diagram of induced voltages: H.V. winding (4)	L.V. winding (5)	Winding connections and relative position of terminals (6)	Phase displacement (Col. 7)	Main group number (8)	Vector group reference number & symbol (9)	Marking of line terminals and vector diagram of induced voltages: H.V. winding (10)	L.V. winding (11)	Winding connections and relative position of terminals (12)
0°	1	11 Yy0						31 Dy1			
		12 Dd0				−30°	3	32 Yd1			
		13 Dz0						33 Yz1			
180°	2	21 Yy6						41 Dy11			
		22 Dd6				+30°	4	42 Yd11			
		23 Dz6						43 Yz11			

FIG. 79 *Illustrating how three-phase transformers may be connected*

lighting to be connected between any phase conductor and the neutral conductor, and large power supplies across the three phase conductors.

Also, it can be seen that no neutral connection is needed on the high tension side and, therefore, the windings may be connected in either a delta or star formation, the star formation being the most suitable for very high voltages. Thus a transformer used for reducing the voltage from 66 kV to 11kV would probably have its primary winding connected in star and its secondary windings in delta, whereas a transformer used for reducing the voltage from 11 kV to 415 V would, in most cases, have its primary winding connected in delta and its secondary windings connected in star.

84. *Describe the conditions which must be satisfied before two three-phase transformers may be operated in parallel.*

Where transformers are required to operate it is necessary that certain requirements should be satisfied. These are:

(1) The polarity of each transformer must be the same.

(2) Each must have the same transformation ratio.

(3) The percentage impedance of each transformer should be of equal value.

(4) There must be the same angle of phase displacement between the respective primary and secondary windings.

(5) The direction of the phase rotation of both transformers must be the same.

These conditions apply equally to both single-phase and three-phase transformers.

(1) This is important because, should the primary windings be cross-connected, short-circuits of very high magnitude could occur in the lines feeding the transformer.

(2) The same transformation ratio ensures that the secondary voltages of the two transformers are the same.

(3) If the impedance of the two secondary windings vary, even if the two output voltages are the same, current will still circulate between the two windings.

(4) Where both the primary and secondary windings are similarly connected (for example delta/delta) the potentials and corresponding terminals on the output side are in the same time-phase sequence. Provided other conditions are satisfied, it is perfectly satisfactory to parallel such a combination and also other combinations such as star/star winding. Referring to Fig. 79, transformers with similar characteristics and of the same Group number may be operated in parallel. It is also possible to arrange the external connections of Group 3 transformers so that they can be paralleled with transformers of similar characteristics in Group 4.

Transformers in Groups 1 and 2 cannot be operated in parallel with one another, without altering the internal connections of one of them, thus altering the group number of the transformer.

(5) The sequence in which the line voltages pass through their cycle is termed phase rotation, and it is necessary that, where transformers have to operate in parallel, the phase rotation of each transformer must be the same.

Where the primary windings are served from the same source, no problems arise from interconnecting the primary windings. Care must be taken to ensure that connections are made to correspondingly lettered terminals on each transformer.

It is possible, however, to connect the two secondary windings so that the polarity is wrong and the phase rotation is in the reverse direction. Before paralleling, therefore, it is necessary to 'phase in' the two transformers. This ensures that the vectors representing the secondary voltages coincided before the interconnections are finally made.

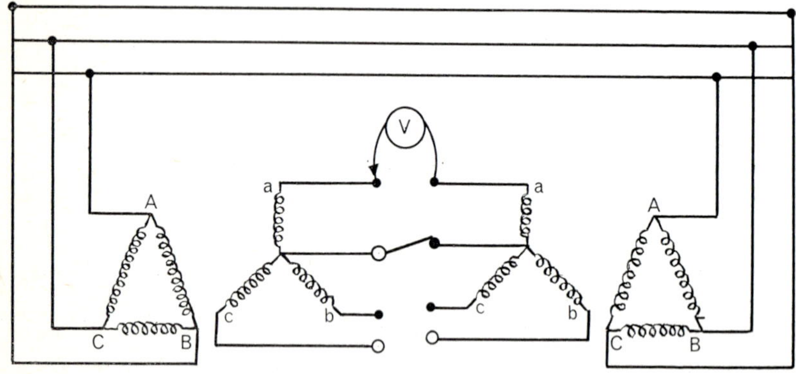

FIG. 80 *Phasing-in two similar three-phase transformers.*

To 'phase in' a pair of similar transformers, the primary windings are connected together, and one terminal of each secondary winding, preferably the neutral conductor (if available) are connected together (Fig. 80). A voltmeter, capable of reading twice the voltage rating is then connected between two terminals which might be connected together. If zero readings are obtained between the respective pairs, it is in order to connect such pairs together. If, however, a reading is obtained between any of the respective pairs of terminals, it indicates that the wrong terminals have been selected, or that the polarities or phase rotations of the two transformers are different.

Where two similar transformers are being connected in parallel, it is not difficult to obtain zero readings on three sets of terminals, but where transformers from different groups have to be paralleled, it might entail considerable testing before paralleling can be satisfactorily achieved.

85. *Describe the switching arrangements where two transformers are required to operate in parallel and state how the maximum flexibility is assured.*

Fig. 81 illustrates the necessary switchgear for operating two transformers in parallel. An essential feature of such an arrangement is that when one transformer is taken out of service for maintenance purposes, the transformer must not be energised from the other transformer, and the isolating equipment used for this purpose should not be able to operate inadvertently while the transformer is

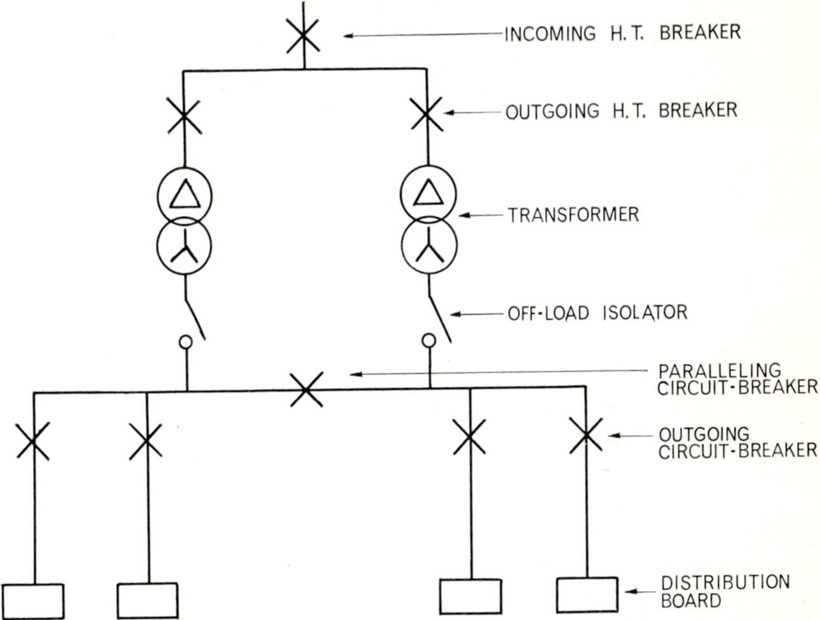

FIG. 81 *Switching arrangement for two transformers in parallel.*

out of service. Since this equipment is used solely for isolation, it need be only of the 'off-load' type (that is, not to be operated when carrying current). A suitable arrangement may consist of two isolators (one for each transformer), operated by handles which may be removed or locked in the 'off' position.

To ensure flexibility, circuit-breakers are incorporated in both the high-voltage and low-voltage equipment. Thus, when all the switchgear is closed, the two transformers are operating in parallel. When one transformer is to be taken out of service, the high-voltage circuit-breaker is opened and also its associated isolating switch. The paralleling switch in the medium-voltage panel would, however, be kept closed so that all the medium voltage equipment is served from the one transformer.

A third method of operating the system is to close all the switchgear except the paralleling circuit-breaker. Each transformer would thus serve separately that part of the medium-voltage switchgear with which it is associated.

86. *Describe how the direction of phase rotation of a three-phase supply may be determined.*

Before instruments are connected to a three-phase supply, it is sometimes necessary to know the phase rotation of a supply to prevent possible damage to them. Although it is possible to use a standard induction motor to determine this, a convenient method is to connect two similar lamps and a capacitor of similar current rating to the three-phase supply, as shown in Fig. 82. The direction of the phase rotation is determined by the brilliancy of the lamps. Because of the presence of the capacitor, one lamp receives approximately 20 per cent of the line

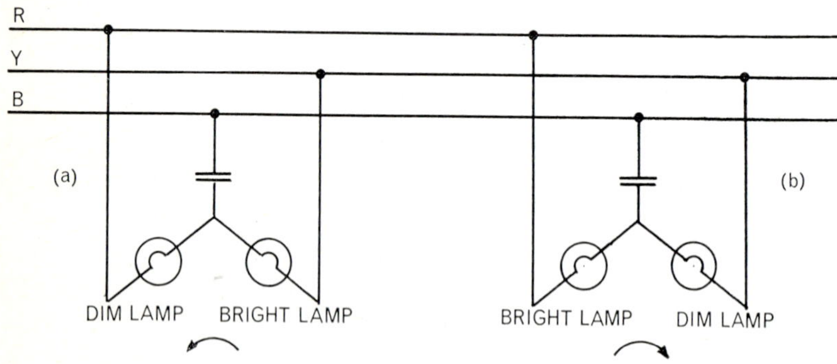

FIG. 82 *Determining the phase rotation of a three-phase supply.*

voltage, while the other lamp receives 80 per cent. Thus there is a variation in the light outputs which is dependent upon the phase rotation. If the arrangement is connected as shown in (Fig. 82(a)), the direction of the phase-rotation is anti-clockwise, the lamp connected to the red busbar being dim and the lamp connected to the yellow busbar being bright. Should the order of the brilliancy of the lamps be reversed (Fig. 82(b)), then the phase rotation is also reversed, being clockwise.

87. *Describe the tests that should be carried out to ensure that the performance of a transformer corresponds to the data on the rating-plate.*

To comply with the data on the rating plate, the following tests should be undertaken:

(1) Ratio and polarity.

(2) No-load current at service voltage and normal frequency.

(3) Open-circuit losses at service voltage and normal frequency.

(4) Load losses at rated current and frequency.

(5) Impedance voltage at rated current and normal frequency on tapping corresponding to service voltage.

(6) Resistance of windings, cold.

(7) Insulation resistance.

(8) High-voltage tests.

To determine the transformation ratio, a comparison of the voltages on the higher-voltage side and the lower-voltage side may be made. Alternatively it may be found by using a special ratiometer transformer (Fig. 83). The no-load current and the open-circuit losses at service voltage and normal frequency may be found from the open-circuit test (question **88**) and the load losses at rated current and frequency from the short-circuit test (question **88**).

The impedance voltage may also be found from the short-circuit test, the voltmeter reading in the primary circuit giving the impedance voltage.

Although not included in the information required on the nameplate, the cold resistance of the windings should be found using either a potentiometer or some form of bridge. The insulation of the windings should be tested using an insulation tester at 1000 V and the high-voltage tests as specified in B.S. 171.

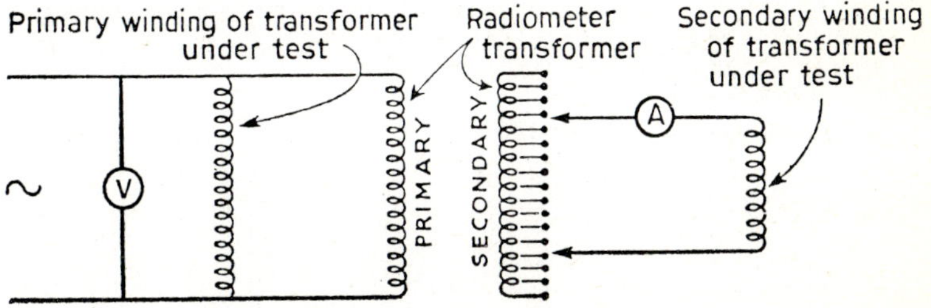

FIG. 83 *Determining the transformation ratio of a transformer.*

88. *Describe the losses that occur in a tranformer. State how they may be determined from the open and short-circuit tests and also how they may be applied in determining the efficiency of the transformer.*

The losses in a transformer may be divided into two groups:

(1) The iron losses.

(2) The copper losses.

The iron losses, however, may be further sub-divided into

(a) The hysteresis loss.

(b) The eddy-current loss.

On standard supplies the iron losses are assumed to be constant irrespective of the load current, but the copper loss varies in proportion to the square of the current.

When a transformer is connected to a 50 Hz supply, the molecules in the iron are also subject to the same alternations. To maintain these alternations means that energy has to be expended and, therefore, the hysteresis loss may be considered as the energy lost in reversing the direction of the magnetism in the material. The higher the number of alternations, the greater is the loss, but as the transformer would normally operate on a standard frequency supply, the loss is assumed to be constant.

When a coil enclosing an iron core is connected to a fluctuating or alternating supply, an e.m.f. is induced not only in the winding but also in the iron core. This gives rise to circulating currents in the iron referred to as eddy currents. If the core was solid, these currents would become excessive, causing heating and power losses with a consequent reduction of efficiency. To minimise the value of the eddy currents, therefore, the iron core is constructed of thin laminations, each lamination being insulated from its neighbour by a thin coating of insulating varnish, so increasing the resistance of the iron to the circulating currents. These losses are also assumed to be constant.

When the two windings are carrying current, there is a heating loss in the windings referred to as the copper losses, which is proportional to the square of the current ($P = I^2R$).

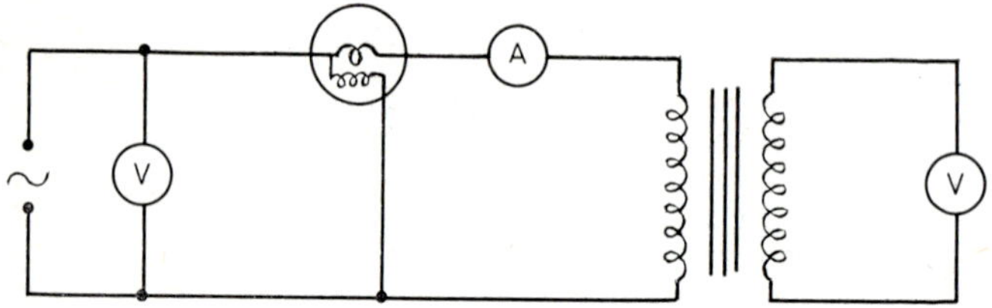

FIG. 84 *Open-circuit test on a transformer.*

To determine the iron losses, the transformer is connected as shown in Fig. 84 (open-circuit test). The ammeter and voltmeters are not strictly necessary for the test, but the voltmeter in the primary circuit indicates that the transformer is operating on its correct voltage, and the ammeter conveniently indicates the no-load current. The voltmeter in the secondary circuit enables the transformation ratio of the transformer to be found.

As the current is very low on no load, the copper losses in the primary winding are negligible; the wattmeter virtually indicates the power necessary to overcome the iron losses in the core. With small transformers, however, it may be necessary to determine the copper losses in the primary winding and subtract them from the value obtained on the wattmeter reading.

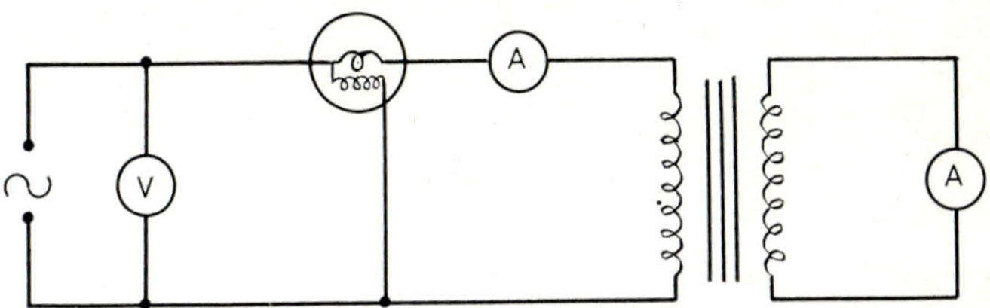

FIG. 85 *Short-circuit test on a transformer.*

To obtain the copper losses, the secondary winding is short-circuited through an ammeter capable of registering the full-load current (Fig. 85). A low, variable voltage is applied to the primary winding, wattmeter, voltmeter and ammeter being connected in the same circuit. The voltage is adjusted, enabling currents at various values up to full load to circulate through both windings.

The voltage necessary to circulate full-load current through the windings when the secondary winding is short-circuited is low, approximately 5 per cent of the normal operational voltage, and as the iron loss is approximately proportional to the square of the magnetic flux (also V), the iron loss may be neglected. The wattmeter, therefore, indicates the copper losses at the various load currents.

Once the iron losses and the copper losses have been determined they may be used to calculate the efficiency of the transformer by applying the following formula

$$\text{Percentage efficiency} = \left(1 - \frac{\text{losses}}{\text{output} + \text{losses}}\right) \times 100$$

or from the formula

$$\text{Percentage efficiency} = \frac{\text{output}}{\text{output} + \text{losses}} \times 100$$

When the output load is at a power factor other than unity, the formula becomes:

$$\text{Percentage efficiency} = \frac{\text{output kVA} \times \cos\phi}{(\text{output kVA} \times \cos\phi) + \text{losses}} \times 100$$

Note that the copper losses at any load can be calculated from one ammeter reading, not necessarily the full-load current.

9

D.C. supplies

Mercury-arc rectifier—selenium metal rectifier—germanium and selenium rectifiers—thyristor.

89. *Describe the construction and operation of a single-phase mercury-arc rectifier.*

The full-wave mercury rectifier (Fig. 86) consists of an evacuated glass bulb containing a pool of mercury (the cathode) in the base: two arms are attached towards the top of the bulb each containing an electrode termed the anode. The anodes, which are normally made of graphites are connected to the outer terminals of a centre-tapped transformer, the centre-tap being taken to the negative output terminal. A further terminal connects the cathode to the positive output terminal.

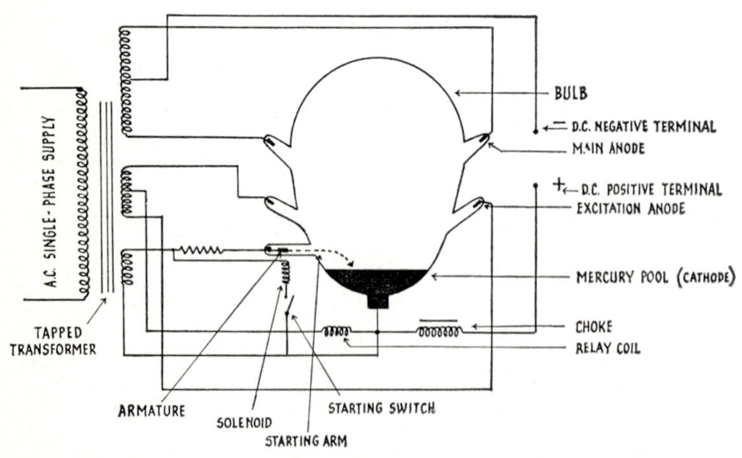

FIG. 86 *Single-phase mercury-arc rectifier.*

Once the discharge is started, it will continue as long as current flows in the external circuit, but if the current is interrupted, the discharge will also cease. To guard against this, small anodes, termed the excitation anodes, are contained in arms also attached to the bulb but much nearer to the cathode than the main anodes.

The excitation anodes are energised from an auxiliary winding on the transformers at a voltage far less than that required for the main anode-cathode circuit. Should the current in the external circuit be interrupted the excitation anodes maintain the discharge.

To initiate the discharge, a special electrode and electromagnet are included in the starting circuit. When the electromagnet is energised, the starting electrode is attracted downwards until it makes contact with the mercury pool. Because this electrode is in parallel with the coil of the electromagnet, when contact is made, the magnet is de-energised so releasing the contact and allowing it to break contact with the pool. An arc is thus formed between the electrode and the pool, causing electrons to be liberated in the vicinity of the arc. The electrons are attracted to whichever of the exciter anodes happens to be positive. As the current in the starting circuit flows through the relay coils, the relay is energised and its contacts remain open for as long as the excitation circuit is operative. The starting circuit is connected between the centre tap and one of the end terminals of the auxiliary winding serving the excitation anodes, and is thus rendered inoperative when the excitation circuit is completed. A limiting resistor is connected in series with the electromagnet and the starting electrode, and protective fuses are also incorporated.

As the electrons leave the mercury pool, they ionise the mercury vapour developed by the 'hot spot' at the surface of the pool. These positive ions bombard the pool to maintain the activity of the 'hot spot'. Thus a continuous supply of electrons is released from the mercury. As the auxiliary winding is centre-tapped, full-wave rectification is obtained even in the excitation circuit.

Adjustable chokes are connected in each anode circuit so that the excitation current can be varied. Surge diverters are normally included in the circuit, which is also protected by fuses.

When the external load is connected, the main circuit is connected in parallel with the excitation circuit, but as the voltage applied to the main anodes is considerably higher than that of the excitation anodes, more electrons are attracted to the former than the latter. The value of the current in the main circuit, however, is determined by the load, so that once the external circuit is established, the main anodes take over the operation of the rectifier.

A choke is normally incorporated in the output circuit so that the output waveform is considerably improved.

To prevent the possibility of backfiring occurring, the condensation of the vapour must keep pace with the vaporization of the mercury, and, therefore, the bulb dimensions are large to enable sufficient vapour to collect and condense before returning as mercury to the pool. To assist in the cooling operation, a fan is mounted beneath the bulb in all but the smallest rectifiers.

As in the excitation circuit, surge diverters are fitted in the main anode leads. These are necessary to prevent damage to the insulation should there be a surge of load current when the rectifier is cold. In some rectifiers, a choke is fitted in the fan circuit to ensure that the fan will only crawl when the rectifier is lightly loaded, and so prevent unnecessary cooling of the bulb. In this case a surge diverter is also necessary.

FIG. 87 *Three-phase mercury-arc rectifier.*

90. *Describe the essential differences in construction and operation between the single-phase and three-phase mercury-arc rectifier and explain with diagrams why the output waveform is smoother for the three-phase rectifier than the single-phase, yet not so smooth as the six-phase rectifier.*

The starting and excitation circuits are the same in the three-phase rectifier as in the single-phase rectifier, but instead of containing two main electrodes, the bulb has three spaced symmetrically round the bulb (Fig. 87). The transformer has three-phase primary and secondary windings. The secondary winding is star-connected, the three free ends being taken to the main anodes while the star point is connected via the load to the cathode.

When the rectifier is in operation, each anode is at a higher positive potential than the other two anodes for one third of the cycle and is therefore conducting. As its potential starts to fall, however, conduction is taken over by the second anode at the point in the cycle when its potential is higher than that of the first anode. This process is repeated with the third anode, so that half-wave rectification is obtained throughout the entire cycle. Thus, a better waveform is obtained with the unsmoothed three-phase rectifier than with the unsmoothed single-phase rectifier, and also a higher value for the average d.c. output is obtained. This can be seen from Fig. 90 which shows a pronounced ripple where the d.c. output is obtained from an unsmoothed single-phase rectifier.

In the six-phase rectifier (Fig. 88), which is virtually a three-phase full-wave rectifier, each phase is operated on both positive and negative half-cycles. This so reduces the ripple that smoothing devices are virtually unnecessary.

91. *Describe the construction and operation of a selenium metal rectifier, and state in what fields it finds its best applications.*

The rectifying properties of this type of rectifier depend upon the difference in the resistance to current flow in each direction: the greater the difference, the more efficient is the rectifier. A good rectifier should offer little resistance in the forward direction and a very high resistance in the reverse direction. As, however, the reverse resistance varies almost inversely as the applied voltage, there is a limit to the voltage which can be applied to each element.

An element of a selenium rectifier (Fig. 89) comprises a plate or disc, usually of steel or aluminium, with a thin layer of selenium deposited on one face. A thin layer of alloy, which may be of lead, tin, bismuth or cadmium, is sprayed on to the selenium to form a counter electrode, the metal disc playing no part in the rectification process. Narrow unsprayed bands are left around holes or any place where there is a danger of contact between the selenium and the alloy.

The elements are mounted on an insulated spindle and are separated by metal

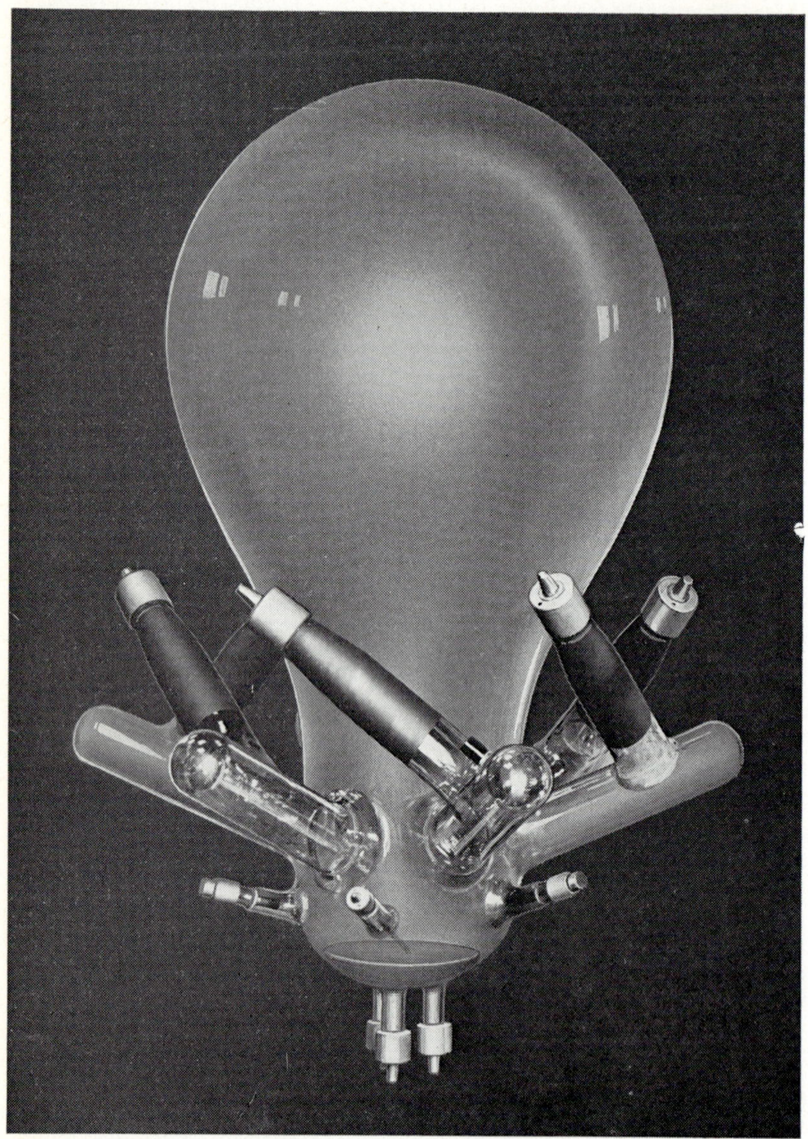

FIG. 88 *Six-phase mercury-arc rectifier. (Courtesy Hackbridge & Hewittic Electric Co. Ltd.)*

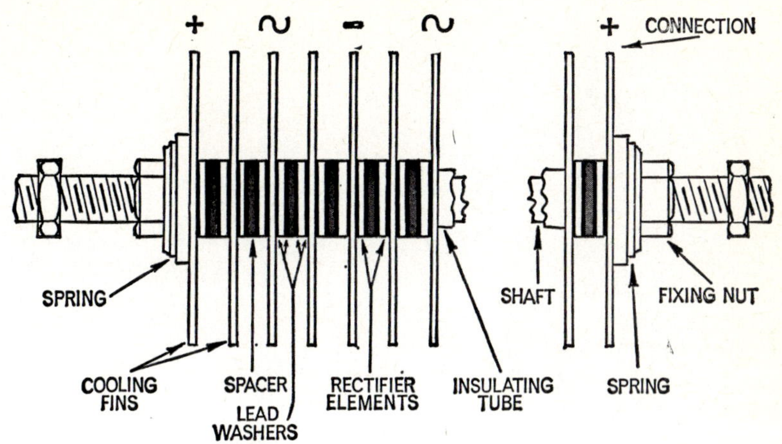

FIG. 89 *Selenium rectifier.*

washers to allow more air to circulate round the elements. Also to help cooling, metal fins may be provided which also serve to anchor the rectifier connections to the external circuit.

The elements will withstand up to 30 V in the reverse direction for normal applications. For higher voltages, elements have to be connected in series. The current rating of each element depends upon the cross-sectional area, and currents of up to 18 A are possible by using plates of large cross-sectional area. For higher currents, the elements may be connected in parallel.

To reduce contact pressure between components, springs are normally fitted at each end of the spindle.

One advantage of the metal rectifier is that the voltage drop in the rectifier is very low, which makes it very suitable for extra-low voltage work such as electroplating installations and battery charging.

It is also very much used for providing direct-current services to cranes, machine tools, lifting machines and magnetic chucks in industrial installations.

92. *Describe, with the aid of diagrams, the meaning of half-wave and full-wave rectification. With connection diagrams, show how full-wave rectification is obtained with* (*a*) *a bi-phase metal rectifier, and* (*b*) *a bridge-connected metal rectifier.*

Fig. 90(a) illustrates the waveform of the alternating current supply. Fig. 90(b) shows how, in half-wave rectification, the negative half of the cycle has been eliminated leaving only the positive half-cycle recurring once every cycle. Fig. 90(c) illustrates how, by suitable connection of the rectifiers, both the positive and negative half-cycles are converted, so resulting in a pulsating uni-directional flow of current in the external circuit.

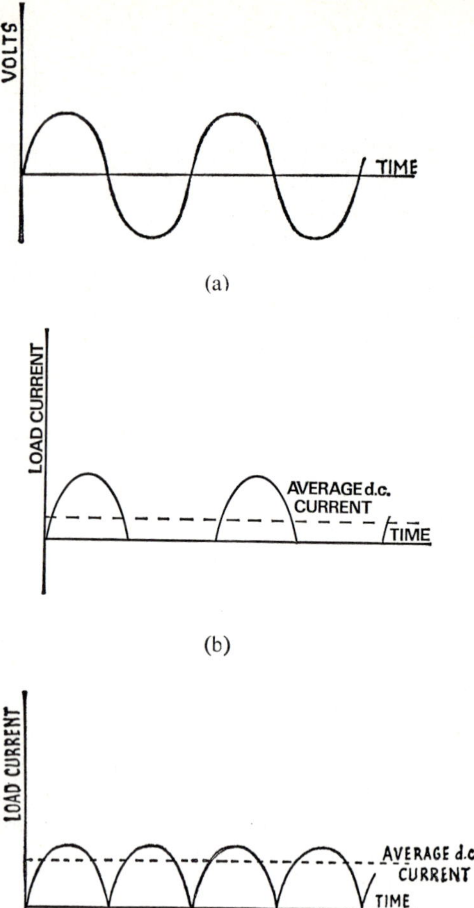

FIG. 90 a, b, c, *Rectified waveforms.*

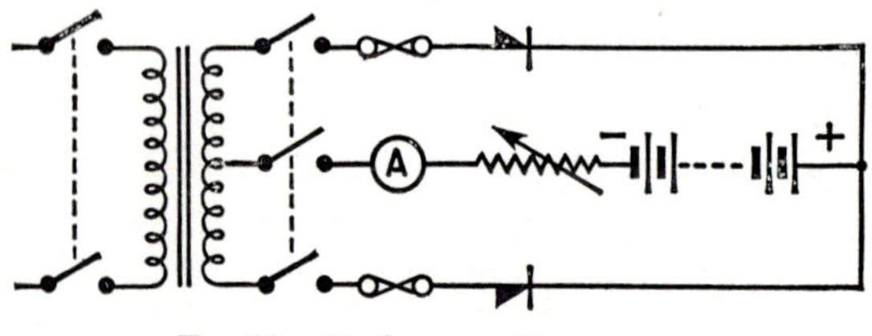

FIG. 91 *Bi-phase rectification.*

For full-wave rectification, two units may be connected in a push-pull arrangement (Fig. 91). Each unit is connected respectively to one end of the secondary winding of a double-wound transformer, the centre point of which is tapped to form one pole of the d.c. output. During one half-cycle, current flows through the top half winding, through the top rectifier and back through the load to the centre tapping. During the second half-cycle, current flows through the lower half winding, through the lower rectifier and through the load again back to the centre tapping on the secondary winding. This method of connection is very suitable where a low d.c. output voltage is required at a high efficiency.

Full-wave rectification may also be obtained by connecting four units in the form of a bridge (Fig. 92). This arrangement requires no transformer, except possibly where voltage adjustment is required, and is capable of attaining very high efficiencies.

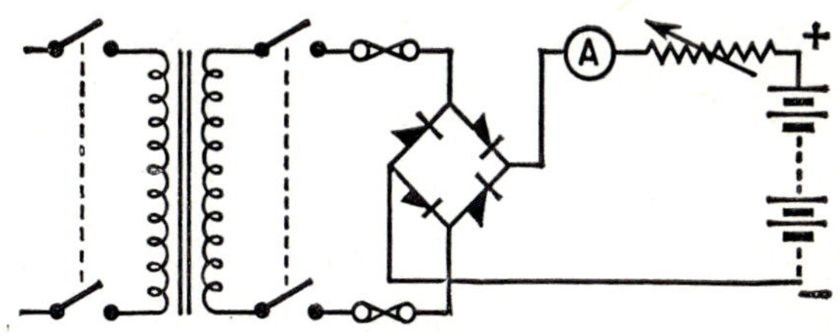

FIG. 92 *Bridge rectification.*

93. *Why is smoothing normally necessary in rectification circuits? Show how it is effected in half-wave and full-wave metal rectification circuits.*

In the unsmoothed rectification circuit, a pronounced ripple is obtained in the output circuit, especially in the single-phase half-wave and full-wave rectifier, and to a lesser degree in the three-phase rectifier. The waveform can be improved by connecting, in the case of a metal rectifier, a reservoir capacitor, or in the case of a mercury-arc rectifier an inductor in the circuit.

When a capacitor is connected as shown in the circuit of Fig. 93, the waveform becomes the one illustrated in Fig. 94. From Fig. 94 it can be seen that the capacitor is charged during part of the positive half-cycle, then as the waveform falls below the capacitor voltage, the capacitor maintains the output voltage at a slightly diminishing level, until, during the next half-cycle, when the positive voltage again exceeds the capacitor voltage.

Although the output d.c. waveform is slightly distorted, the rectifier has an improved waveform and a higher output than that of the unsmoothed rectifier.

In the smoothed full-wave rectifier circuit, the waveform of which is shown in Fig. 95, the capacitor is charged every half-cycle, giving a better waveform than the smoothed half-wave rectifier.

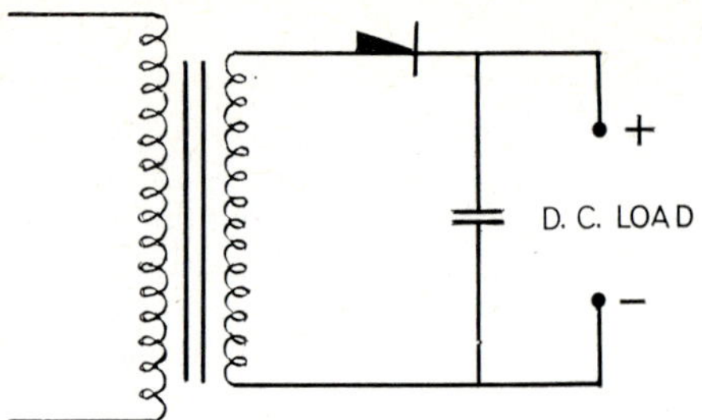

FIG. 93 *Smoothed half-wave rectification circuit.*

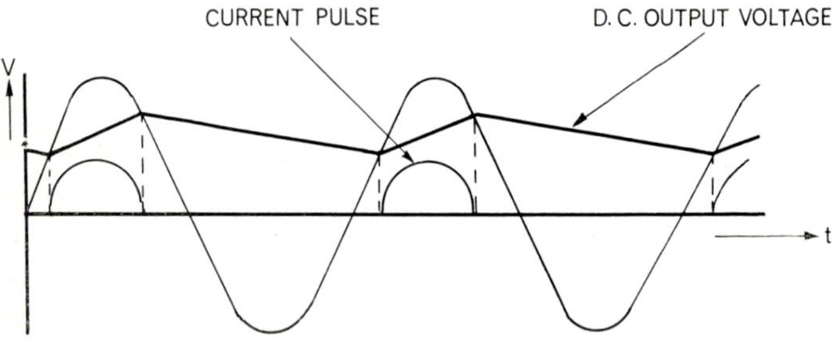

FIG. 94 *Waveform for smoothed half-wave rectification.*

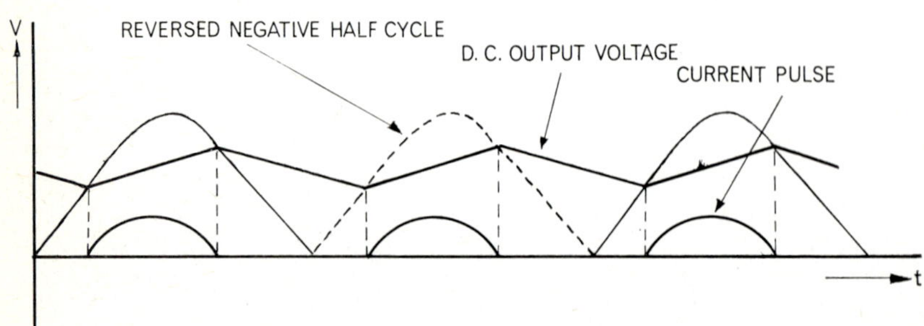

FIG. 95 *Waveform for smoothed full-wave rectification.*

94. *Describe the precautions to be observed in the installation and maintenance of metal rectifiers.*

If the rectifier is to give satisfactory service throughout its life, the following precautions should be taken:

The rectifier must not be loaded beyond its normal rated output. A further precaution is to ensure that the input voltage at no load does not exceed the maximum working voltage of the rectifier.

The rectifier should be installed in a well-ventilated area, remote from other units which dissipate heat and would raise the ambient temperature. Where the ambient temperature is high, it is necessary to de-rate the rectifier.

Where the rectifier is exposed to damp or corrosion, it should be given an anti-corrosive painting by the manufacturer before being despatched. Where conditions are particularly severe, it is advisable to install oil-immersed rectifiers.

At no time should the rectifier be painted as this may be harmful to its physical properties.

It is usual to institute a periodic check of all connections, etc., but no special maintenance is required, other than occasionally blowing out any accumulation of dust or dirt.

Metal rectifiers are often used on circuits carrying very heavy currents at extra-low voltage, such as in electroplating and electrodeposition. To avoid excessive voltage drop in the cable and switchgear, the rectifier should be installed as near as is practicable to the plating vat. Since corrosive atmospheres are normally associated with electroplating, it is advisable to install oil-immersed rectifiers.

95. *Describe the construction and operation of germanium and silicon rectifiers and state the advantages they have over other forms of rectifiers.*

Both germanium and silicon belong to the class of materials known as semiconductors, the conductivity lying between that of metals and insulators, and in their natural form they do not possess the property of rectification. If, however, a small amount of impurity (arsenic for germanium and phosphorus for silicon) is introduced into a perfect lattice of either material, the structure of the lattice is altered and there is a surplus of free electrons. The material treated is referred to as an *n-type* material; in this form current is able to flow readily through the crystal.

Introducing other impurities into the crystal (indium for germanium, boron for silicon) produces a deficiency of free electrons, leaving what are termed 'holes'. In this form the material is classed as a *p-type* material.

If a piece of germanium or silicon contains both *p*- and *n*- type materials, a potential barrier is established at the junction which prevents the passage of electrons or holes. If now, with the material connected to a source of e.m.f., the *n*-type material is made positive with respect to the *p*-type material, the barrier layer is increased and virtually no current can flow through the material. If the *p*-type is made positive with respect to the *n*-type, the potential of the barrier layer is neutralised and the current is limited only by the resistance of the material and external circuit.

FIG. 96 *Germanium and silicon diodes.*

These rectifiers can be manufactured with very low cubic capacity in relation to their power output, Fig. 96 illustrating a diode of each type. The diodes are shown mounted on a suitable base which accounts for a large proportion of the cubic capacity, but even so the current density is in the order of 0·4 A/mm² for germanium and 1·2 A/mm² for silicon elements.

The germanium rectifier will withstand up to 1000 V and the silicon rectifier, 1500 V, but because voltage surges can produce abnormally high currents harmful to the rectifier, the safe working voltages are in the order of 250 V for germanium and 350 V for the silicon rectifier. The voltage drops in the rectifiers are in the order of 0·5 V for germanium and 1 V for silicon. Because the operating voltages are high, very high efficiency values are obtained.

Germanium rectifiers operate satisfactorily in temperatures up to 50° C. Above this temperature they must be de-rated and, in any event, should not be operated in temperatures above 80° C. The silicon rectifier operates satisfactorily in temperature up to 180° C, and normally de-rating is unnecessary.

Both types of rectifiers require protection against normal overload and short-circuit. Where normal overloads are rare, h.b.c. fuses should be satisfactory, but where overloads are rather frequent, automatic protection is preferable. For short-circuit protection, a reactor in the a.c. supply will normally limit the short-circuit current to eight to ten times the full-load current.

The diodes may be connected in series to obtain higher voltage outputs, and in parallel for higher current putputs. Any of the circuits shown in Fig. 91 or 92 may be used to obtain half-wave or full-wave rectification.

The advantage of the germanium rectifier is its high efficiency in the lower voltage ranges, and since it will operate at a higher reverse voltage than the selenium rectifier, fewer elements would be required for the same output voltage. The most suitable application, therefore, for the germanium rectifiers falls within the 8 V to 150 V range, especially where high currents are required.

The silicon rectifier is now being used for a variety of electrical devices ranging from electronic equipment to the largest power units. It is far more suitable than the germanium rectifier where the ambient temperature is liable to be abnormally high, and since it will also withstand high reverse voltages, it is preferable where high voltage outputs are required.

96. *Describe the construction and operation of the thyristor (controlled rectifier) and briefly state its advantages and applications.*

Basically, the thyristor is a rectifier which, in some ways, resembles the germanium and silicon rectifier. The difference is that, in the diode, current is blocked in the reverse direction, whereas in the thyristor, current can also be blocked in the forward direction (that is when the anode is positive with respect to the cathode). In this respect, it resembles the grid-controlled mercury-arc rectifier and thyratron valve, in that there is no current in the device until the gate is sufficiently positive, when the current is no longer inhibited and the thyristor becomes an ordinary rectifying device.

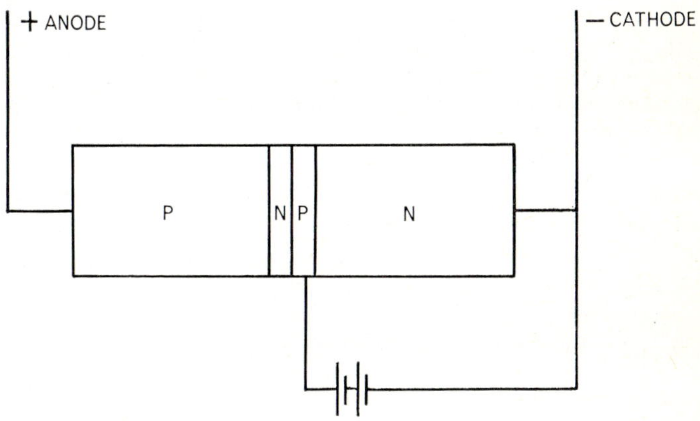

FIG. 97 *Illustrating the operation of a thyristor.*

The construction of the thyristor differs from that of the silicon diode in that a thin wafer of silicon is embedded in the former (Fig. 97) which is referred to as the gate. This wafer is processed and refined into a semiconductor with positive and negative layers. The important point to remember is that the potential applied to the gate determines the current that will flow through the device. Hence, in addition to the anode and cathode connections, a third connection must be taken to the gate.

It has already been said that once the device has been 'triggered' it behaves like a normal diode: as in the grid-controlled rectifier, removal of the positive potential at the gate will not inhibit the flow of current. Only when the anode voltage has fallen to zero, or the anode current is below a certain limit (referred to as the sustaining current) will the device cease to conduct.

This sustaining current is important in that, if the current in the device fails to rise above this value, it will cease altogether when the gate pulse is removed. Normally this pulse lasts only for a few milliseconds, so that it is possible, if the load is highly inductive, for the current to lag so far behind the voltage that it will not rise sufficiently for the sustained value to be exceeded, owing to the very short duration of the pulse. The interval during which the device is triggered would therefore have to be increased. In effect, the earlier or later in the positive half-cycle the gate is triggered, the higher or lower will be the average output voltage. Thus, by this method of phase controlling the current in the device, the energy output may also be controlled between zero and its maximum value. As the energy consumed by the gate pulse is very low, it constitutes a most effective means of controlling the power input.

The voltage drop in a thyristor is very low, in the order of 1·8 V, which is below that obtained with the mercury rectifier or thyratron valve. The small voltage drop, therefore, produces a small power loss in the device and as the internal heating depends upon the power loss and time, it follows that quite small units can serve for a.c./d.c. conversions at substantial power levels.

Thus, one of its main advantages is the small physical space it requires compared to that required by the mercury-arc rectifier of comparable rating. Another advantage is that the thyristor operates much faster under control, and its conversion efficiency is also much higher than either the mercury-arc rectifier or the thyratron valve.

Thyristors are being increasingly used for controlling the speed of d.c. motors, and when connected in a bridge formation, they considerably reduce the size and weight of the whole equipment. Other applications include: energy regulators (possibly one of the most simple applications if the load is not too reactive); inversion where thyristors provide a relatively simple means of obtaining an a.c. output from a d.c. input; the chopper control of d.c. motors, The thyristor may also be used from a d.c. source provided the supply is intermittent.

10

Power factor

Benefits of power factor correction—methods of application—automatic power factor correction.

97. *What are the benefits of power factor correction?*

Power factor correction benefits the Generating Boards, the Supply Authorities and the consumer. The benefits to the consumer are:

(1) The power losses in his cables and switchgear, etc., are reduced.

(2) It enables him to install more useful load for given sizes of cable and equipment.

(3) It materially reduces his overall electricity bill.

It should be noted that the cost of installing corrective equipment is normally repaid in the saving in electricity charges within a period of eighteen months to two years.

The benefit to the Generating Boards and Supply Authorities is that, if the load were left uncorrected, far larger generators, transformers, switchgear and cables would be necessary. If the Authorities had to apply the correction themselves for all the low lagging power factor loads connected to their supply, very large banks of corrective equipment would have to be installed at a relatively high cost. Since this cost would have to be passed on to all consumers, this would be very unfair to those consumers who had raised the power factor of their own installation to a high value by installing corrective equipment.

The Supply Authorities therefore endeavour to make those consumers whose power factor is low, improve it by imposing penalty clauses if the power factor falls below a predetermined value, normally 0.9.

Two tariffs by which the consumer is penalised for a low power factor are:

(1) Payment of a fixed rate for each kVA of maximum demand plus a small charge per unit indicated on the kWh meter. An important point to note is that the demand-indication needle registers the maximum kVA taken by the instal-

lation for that particular month. Leaving the correction equipment out of circuit for only a short period might lead to an excessive value being registered on the meter.

(2) A charge for every kW registered on the demand meter, plus a small unit charge, the demand charge being increased by 1 per cent for every 0·01 by which the power factor falls below 0·9. Should the power factor fall to, say, 0·75, the demand charge would be increased by 15 per cent.

98. *What methods may be employed for installing power factor correction equipment?*

(1) *Across the busbars of section boards feeding groups of plant.* This method is very suitable where the load variation depends upon the operating hours, especially where the load consists of a large number of small motors, because, if any section is disconnected for any length of time, the power factor for the remainder of the installation is not materially affected. Also the effect is not so severe if the correction equipment is left in circuit with the load disconnected.

The capacitor would normally be mounted adjacent to the section board and be connected to the busbars through a switch-fuse, preferably with h.b.c. fuses. Large capacitors are preferably controlled and protected by a circuit-breaker. Although the individual circuits beyond the fuseboard would not have correction applied to them, all the circuits and switchgear between the section board and the alternator would have their currents reduced, thus reducing the power losses in those parts of the installation.

(2) *Across the busbars of the medium-voltage switchboard.* Where bulk correction is applied to an entire, large installation or even to large sections of of the installation, it is almost essential to install automatic equipment. Where non-automatic bulk correction is appled to such installations, it is very difficult to determine the correct value to apply owing to the possibility of wide fluctuations of load. If the assessment is based on the maximum load, the installation may possibly take a leading power factor when the factory is running lightly loaded. If the assessment is based on a mainly light load, the power factor will be low when the installation is running heavily loaded. Also, if the load is switched off and the capacitor is inadvertently left in circuit, the installation will be operating with a very low leading power factor.

(3) *Individual correction at the stator terminals of induction motors.* This method is seen at its best advantage when it is supplementary to other methods of correction. Thus, bulk correction may cater for the minimum loading on the installation, while the remainder of the required correction is divided into small units connected to individual motors. Variation of load on any motor has little effect on the remainder of the installation, while shutting down the motor ensures that the capacitor, which may be connected across the terminals of the motor or across the starter, is disconnected. By connecting the capacitors across

the motor terminals, the current is reduced in all parts of the circuit right from the motor terminals back to the alternator, with a consequent reduction in the power losses. In practice, for most installations, a small amount of bulk correction, plus individual correction for the large induction motors and discharge-lamp circuits, is the most simple and reliable method of power factor correction.

Where capacitors are connected across the terminals of the motor, it is unnecessary to provide separate control gear, but, if necessary, fuses may be incorporated in the capacitor for isolation purposes.

(4) *Individual correction of discharge lighting circuits.* This method is similar to the last method but is normally used on single-phase circuits. Owing to the inherent low power factor of uncorrected inductor-type discharge-lamp circuits, power factor correction is normally applied at each lamp individually. Although the cost per fitting is increased, the method does ensure that, where a lighting installation consists of inductor-operated discharge lamps, the power factor of the installation is kept at a high value.

Also, many discharge lamps are fitted in small installations which would not normally install separate corrective equipment. These installations comprise a a large proportion of the number of consumers connected to the public supplies. Individual correction does ensure that the Supply Authority's equipment and cables are not subjected to any increase of current due to the addition of uncorrected small installations with low power factors.

99. *How may bulk power factor correction be automatically applied?*

The most common method of automatic control is by the installation of a reactive volt-ampere sensitive relay which maintains the reactive current within pre-determined limits, irrespective of the load conditions. This is achieved by the relay operating a contactor or contactors which switch in capacitors as required. The full operation is as follows.

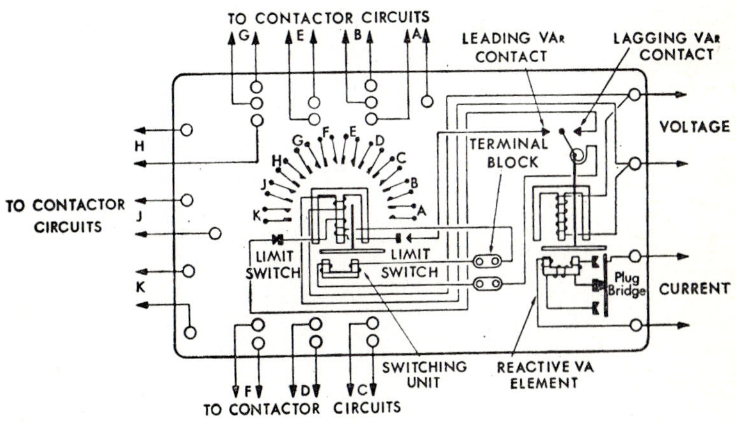

FIG. 98 *Automatic power factor correction.*

When the power factor of the installation is lagging, the lagging contacts in the relay close, so that the motor-driven controller rotates in one direction, operating the mercury switches (A, B, C, D etc., in Fig. 98) successively and bringing into circuit, step-by-step, banks of capacitors until the lagging current in the circuit is reduced to the required amount or until all the capacitors have been called into service. When the load is reduced and the leading reactive current starts to exceed the lagging reactive current, the leading contact in the relay is made and the motor-driven controller rotates in the opposite direction, opening the appropriate number of mercury switches until once again, the power factor is stabilised at the required value.

The mercury switches are of the heavy duty type enabling them to control the capacitor contactors without any intermediate relay, and limit switches are provided to prevent overdriving of the camshaft in either direction. The driving motor is solidly coupled to the camshaft and special high-lift cams are fitted for actuating the mercury switches.

The relay operates similarly to the watt-hour meter, but as the torque is proportional to the reactive power, not the power, it is necessary that the current coil should be connected to the phase other than the two to which the voltage coil is connected.

11

Instruments

Moving-coil instrument—moving-iron instrument—dynamometer instrument—potentiometer—Wheatstone bridge—slide-wire bridge—energy meter—instrument transformers—measurement of power.

100. *Describe the construction and operation of a moving-coil instrument and state why is it unsuitable for a.c. supplies. How may its range be extended?*

The instrument shown in Fig. 99 comprises a field system of a central magnet which is surrounded by a soft-iron yoke, specially shaped iron pole shoes ensuring a uniform flux for an evenly divided indicator scale. The moving system consists of a coil of insulated, very small diameter wire wound on an aluminium former, which is suspended in jewelled bearings. The former is suspended in the annular space between the central magnet and the yoke and a pointer is attached to the spindle, so that the movement of the moving system is indicated on a graduated scale. Two hairsprings, wound in opposite directions, are also attached to the moving system, In addition to providing the controlling torque, they also serve to connect the coil to the supply. One spring is also attached to a zero adjustment lever.

When current is flowing through the coil, the magnetic field established by it interacts with the magnetic field of the permanent magnet so causing the moving system to rotate, the deflection being proportional to the current flowing through the coil. The controlling torque necessary to prevent the moving system from rotating too far is provided by the hair-springs, the needle coming to rest when the controlling torque and the deflecting torque are equal. Since the rotation is proportional to the current in the coil, the scale is evenly divided. A damping torque is essential to prevent the needle from oscillating about its mean position and this is provided by constructing the former of aluminium. As the moving system rotates in the magnetic field, eddy currents are induced in the former which rapidly brings the moving system to rest. Since the magnetic field is

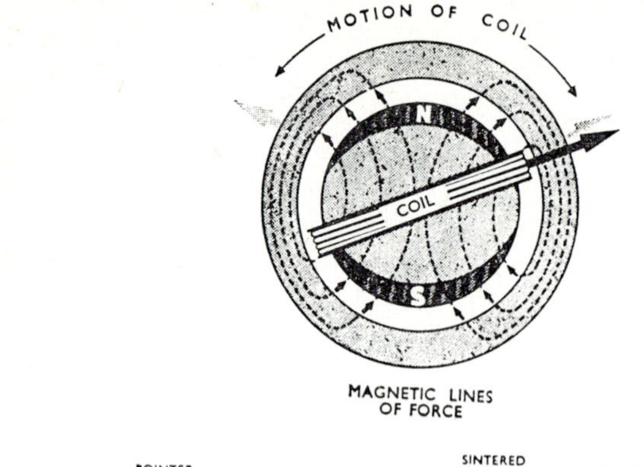

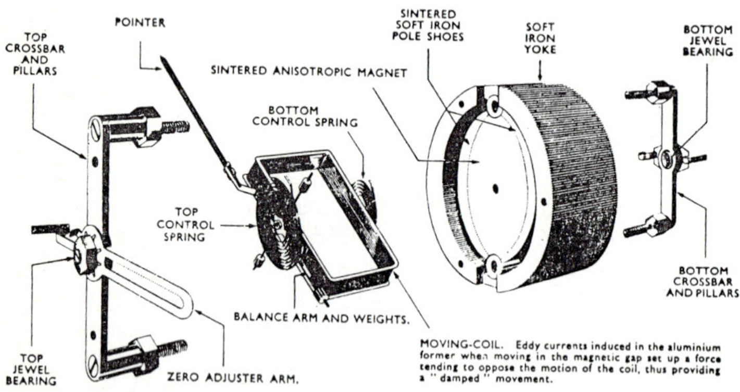

FIG. 99 *Moving-coil instrument.* (*Courtesy Crompton Parkinson Ltd.*)

produced by a permanent magnet, any reverse flow of current through the coil would cause the moving system to endeavour to rotate in the reverse direction, hence the instrument is only suitable for operating on d.c. supplies.

Because of its delicate movement, however, the instrument has a very sensitive operation, and therefore it is often used in conjunction with a rectifier for the measurement of a.c. currents and voltages.

As the current in the coil has to flow through the hairsprings, its value is limited by the amount the springs can safely carry. A normal instrument has an operating current in the order of 15 mA and a coil resistance of 5 Ω, so that the potential drop across it for full-scale deflection is 75 mV.

For the instrument to read normal supply voltages and currents, resistors have to be incorporated to limit the current to a safe value. In the moving-coil voltmeter, the resistance is connected in series with the instrument, its value being such that, although the potential drop across the instrument is 75 mV for full-scale deflection, the scale reading is calibrated to read the voltage of the supply across which the voltmeter is connected.

By incorporating a number of graded resistances the one instrument may be used for the measurement of supplies at various voltages.

The ammeter has to be connected in series with the load, and therefore the potential drop across the instrument and its associated resistor (termed the shunt) must be very low, so that little voltage is lost across the load. The shunt, therefore, has a very low resistance value and is connected in series with the load, while the instrument is connected in parallel with the shunt.

Thus, if the shunt is calibrated correctly, irrespective of the current flowing through it, only 15 mA would flow through the meter for full-scale deflection, the remainder flowing through the shunt. If a number of shunts are graded so that, irrespective of the current flowing through them, the current flowing through the meter is limited to 15 mA, then the one instrument may be used for the measurement of current of various rated circuits.

As the shunt has a very low value of resistance, its value must not be affected by changes of temperature (that is the temperature coefficient of resistance must be negligible), hence it is usually constructed of materials such as Manganin. Also, where leaved shunts are used for measuring large currents, the leaves should be mounted vertically to allow a better dissipation of heat.

By combining various shunts and series resistors, sometimes referred to as swamps, the one instrument (Fig. 100) may be combined to read voltages and currents over several ranges, and if a rectifier is included, for both a.c. and d.c. supplies.

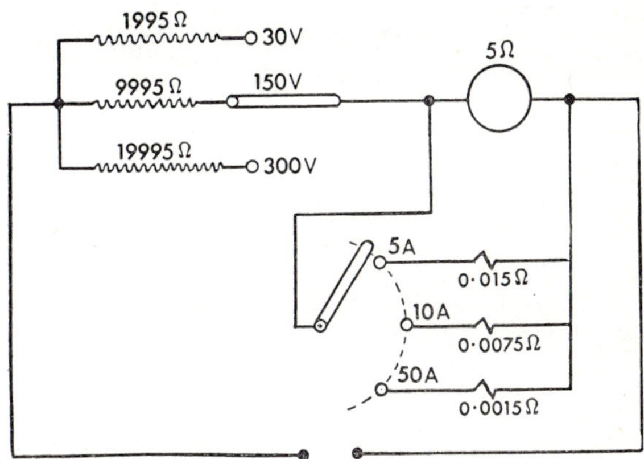

FIG. 100 *Extension of the range of a moving-coil instrument.*

101. *Describe the construction and operation of a repulsion-type moving-iron instrument and state its advantages and disadvantages.*

The repulsion-type moving-iron meter comprises a circular coil which carries the current to be measured and two pieces of soft iron, one fixed to the coil and one attached to the moving system (Fig. 101). When current is flowing

F*

through the coil, both pieces of iron are similarly magnetised so that a repulsive force is exerted between them. As this repulsive force produces a deflecting torque which is proportional to the square of the r.m.s. current, the scale is unevenly graduated being cramped at the beginning and open towards the end of the scale. The scale, however, can be somewhat improved by altering the design of the pieces of iron.

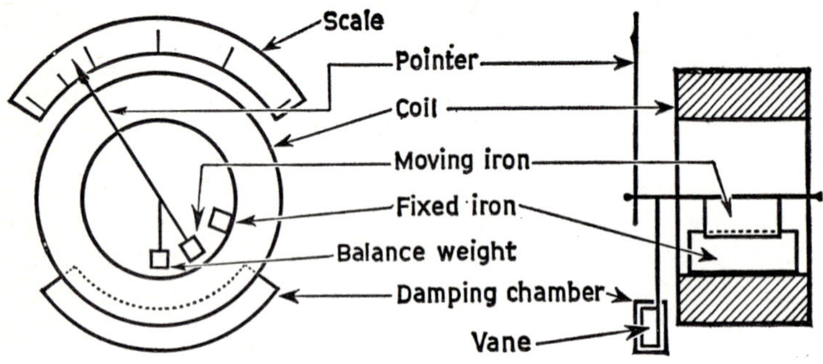

FIG. 101 *Repulsion-type moving-iron instrument.*

The controlling torque may be produced by attaching weights to the moving system, which limits the position in which the instrument can be used, or spring control may be adopted.

Air damping is normally provided. Because any form of electromagnetic damping necessitates a permanent magnet, the field would tend to affect the accuracy of the reading.

Since reversing the current flow reverses the polarity of both pieces of iron, the force is maintained in the same direction, and the instrument may be used for measuring both a.c. and d.c. supplies.

Within limits, the meter may be used for measuring several values of current by tapping the coil at selected points, but for currents in the higher ranges, instrument transformers would have to be incorporated, shunts being unsuitable on a.c. supplies.

The same movement may be used for a *voltmeter*, but, in this case, the required number of ampere-turns is obtained by winding the coil with a large number of turns of comparatively fine wire to keep the operating current low. Also, to minimise temperature errors, a swamping resistance is included, constructed of Manganin wire of such value that it forms the major part of the total resistance. The swamping resistance may also be used to extend the range of the voltmeter.

The advantages of the moving-iron instrument are:

(1) It is very robust in design and construction.
(2) It is relatively inexpensive.
(3) It can be used for measuring both a.c. and d.c. supplies.

The principal disadvantages are:

(1) It is liable to be affected by stray magnetic fields, although this may be

minimised by screening the movement (that is housing it in an iron case).

(2) It is subject to hysteresis error when used on d.c. supplies (the instrument reads higher with decreasing than with increasing values of current.) This may be partly compensated by careful design of the iron pieces or by using materials with a low hysteresis loss (for example, Mumetal).

(3) Variations of frequency may appreciably affect the accuracy of the instruments, particularly voltmeters, because of the inductance of the coil. On standard supply frequencies, this is not of great importance, but its effect may be minimised by making the resistance of the coil high compared with its reactance. If necessary, the use of a capacitor can make the instrument virtually independent of frequency.

(4) Moving iron voltmeters are also subject to temperature error, which is partly combated by using a series resistance with a negligible temperature coefficient.

102. *Describe the construction and operation of a dynamometer instrument and state how it would be connected to read the power taken on a single-phase supply.*

The dynamometer instrument (Fig. 102(a)) comprises a fixed coil, normally divided into two sections, and a moving coil, lying within the fixed coil, supported in jewelled bearings. The moving coil is also attached to the needle which moves over a graduated scale. As in the moving-coil instrument, the controlling torque is produced by hairsprings which also serve to lead current into and out of the moving coil.

FIG. 102 (a) *Dynamometer instrument.* (*Courtesy Crompton Parkinson Ltd.*)

The fixed coil produces a field of almost uniform density, and the deflecting torque results from the interaction between this magnetic field and that produced by the current flowing through the moving coil, the deflecting torque being proportional to the products of the currents in the two coils. In ammeters, it is normal to connect the fixed and moving coils in parallel, whereas in voltmeters they are connected in series. In each case, the deflection is proportional to the square of the current so that the scale is uneven.

If the current is reversed, its direction is reversed in both the fixed and moving coils, so that the direction of torque remains unchanged. Consequently, the instrument can be used on both a.c. and d.c. supplies. The coils are normally air-cored, to avoid errors that would be produced by using iron cores, and since the presence of a magnet is liable to affect the accuracy of the instrument, air damping is adopted. Also, since the operational field is weak, the instrument requires shielding from stray fields.

The dynamometer instrument finds its best application, however, when used to measure the power absorbed in a circuit. If the fixed coil is so arranged that it carries all or part of the current in the circuit, while the moving coil is connected in series with a non-inductive resistance of relatively high value across the supply potential, so that the current in it is in phase with the applied voltage, the deflecting torque is proportional to the product of the instantaneous currents in both coils, and hence the power in the circuit. In the case of the wattmeter, the scale reading is substantially evenly divided.

Fig. 102(b) shows how the wattmeter is connected to read the power taken by a single-phase circuit, and also, how by including a voltmeter and ammeter, the power factor of the circuit may also be calculated.

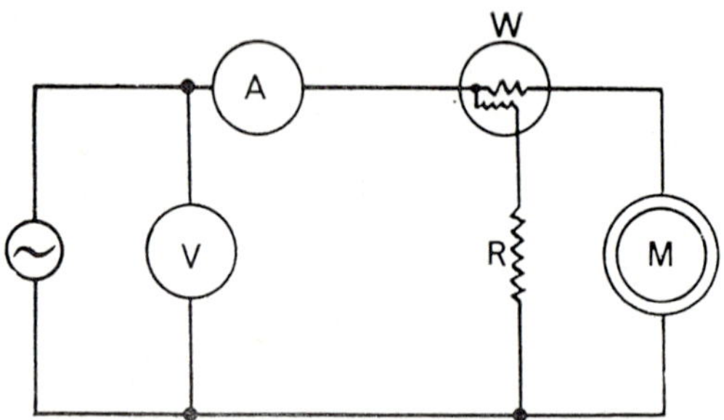

FIG. 102 (b) *Connections for an ammeter, voltmeter and wattmeter in a single-phase circuit.*

103. *Describe the construction and operation of the simple potentiometer, and state for what purposes the commercial form is used in industry.*

The potentiometer is one of the most useful of instruments for the accurate measurement of current, p.d. and resistance. In its simplest form (Fig. 103) it consists of a wire of uniform cross-section, alongside of which is a graduated scale. An accumulator is connected across the full length of the wire. Assuming that the e.m.f. of an unknown cell is required, a standard cell, such as a Weston cell, is first connected in series with a galvanometer to a sliding contact on the wire, which is adjusted until a zero reading is obtained on the galvanometer. The length l_1 is noted, and the p.d. across this length has the same value as that of the standard cell.

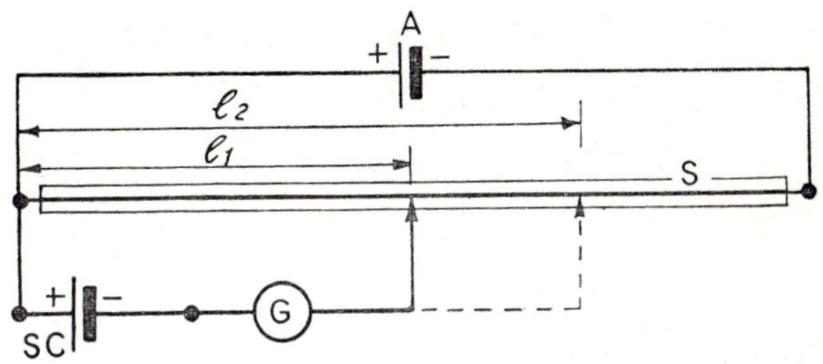

FIG. 103 *The simple potentiometer.*

The standard cell is replaced by the cell of unknown value, and the sliding contact is readjusted until a zero reading is again obtained on the galvanometer. The length l_2 is noted, and as the p.d. across this length has the same value as that of the cell of unknown value, the following equation is true:

$$\frac{E_1}{E_2} = \frac{l_1}{l_2}$$

where E_1 and E_2 are the e.m.f.s of the standard and unknown cells, respectively.

Hence
$$E_1 = \frac{E_2 \times l_1}{l_2}$$

or
$$= \frac{1{\cdot}0186 \times l_1}{l_2}$$

assuming the standard cell is a Weston cell.

It is very difficult to obtain quick and accurate measurements with the simple potentiometer, but the direct-reading type fulfils these requirements. The principle is the same, but the settings are much more easily effected, and, therefore, the commercial instrument is used extensively for accurate calibration of voltmeters, ammeters and resistors. Fig. 104 illustrates a direct-reading potentiometer set for the calibration of an ammeter.

104. *Describe the construction of the Wheatstone bridge and state how it may be used to find the value of an unknown resistance*

The Wheatstone bridge comprises a network of four resistors, two of known value, one variable and one resistor of unknown value, a galvanometer, battery and the necessary switches. The circuit is shown in Fig. 105.

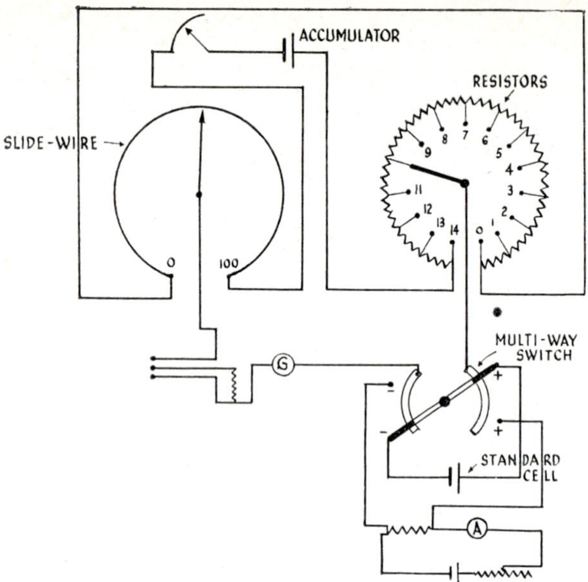

FIG. 104 *Calibration of an ammeter using a direct reading potentiometer.*

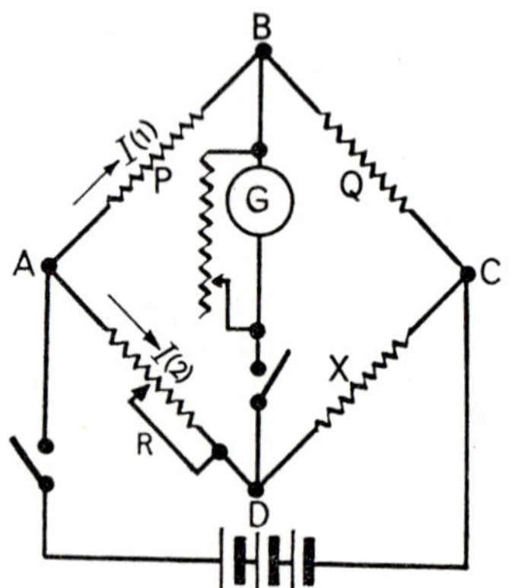

FIG. 105 *The Wheatstone bridge.*

The principle of operation is that, to find the value of the unknown resistance *X*, the bridge has to be 'balanced' by adjusting the variable resistor *R* until the galvanometer reading is zero. Thus when balance occurs, the following equations are true:

$$\text{p.d. across } P = \text{p.d. across } R \tag{1}$$

and

$$\text{p.d. across } Q = \text{p.d. across } X \tag{2}$$

substituting IR values and dividing equation (1) by equation (2)

$$\frac{I_1 \times P}{I_1 \times Q} = \frac{I_2 \times R}{I_2 \times X}$$

where P, Q, R and X are the resistance values,

therefore,

$$\frac{P}{Q} = \frac{R}{X}$$

cross-multiplying

$$X = \frac{R \times Q}{P}$$

Although, during the test, the values of P and Q are fixed, by arranging for different ratios to be substituted, very low and very high values of the unknown resistor may be calculated, depending upon which direction the ratios of the known resistors are in. Thus, if the resistance of Q is high compared to that of P, then it enables resistors of high value to be calculated, whereas, if the resistance Q is low to that of P, then resistances of very low value may be calculated, that is the scope of the bridge can be very much widened by carefully selecting the values of the known resistors.

The importance of the Wheatstone bridge method of determining the values of resistance lies in the fact that it is a null method, that is although current is being drawn from the battery, in the essential part of the circuit the current is zero for balanced conditions.

105. *Describe the ratiometer principle and state how it is applied in the insulation tester.*

This principle is used in many types of instrument. Basically it consists of mounting two coils (A and B in Fig. 106) at a fixed angle to each other and suspending them in a uniform magnetic field. If the two coils are so wound that the torque of one coil opposes that of the other, the resultant torque exerted on the moving system is dependent upon the ratio of the currents in the two coils. Thus if the current in coil A is greater than that in coil B, the coil rotates clockwise, whereas, if the current in coil B is greater than that in coil A, the coils will rotate anti-clockwise. No spring control is necessary, the moving system reaching its steady value when the two torques are equal.

One form of insulation tester (Fig. 107) consists of a d.c. generator or a.c. generator with rectifier, and a moving coil instrument which operates on the ratiometer principle. One coil (termed the control or voltage coil) is connected in series with a resistor directly across the generator, whereas the other coil

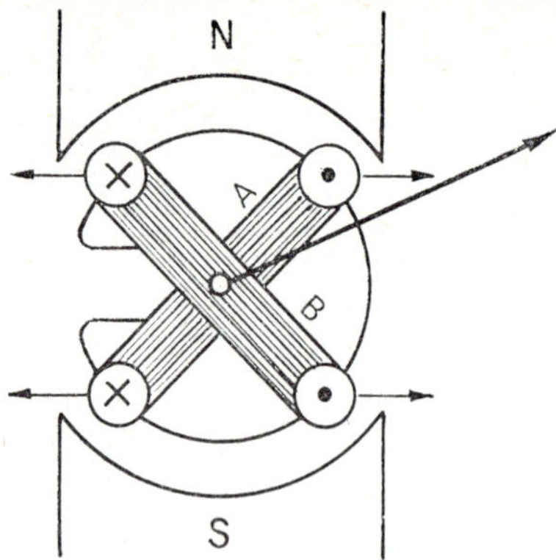

FIG. 106 *The ratiometer principle.*

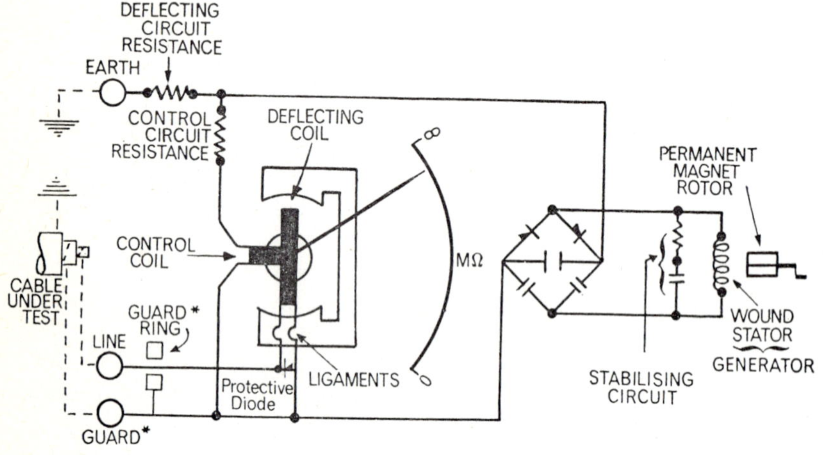

FIG. 107 *The combined continuity and insulation tester.*

(the deflecting or current coil) is connected in series with a test resistor, also across the generator. The two circuits are therefore connected in parallel across the generator.

When the test terminals are open-circuited, no current flows through the deflecting coil, and the torque generated by the control coil causes the moving system to rotate until the needle points to the infinity mark on the scale. When the test terminals are short-circuited, maximum current flows through the deflecting coil. The torque generated by this coil overcomes the torque of the control coil so that the moving system rotates in the opposite direction until the needle points to the zero mark. The instrument may therefore be calibrated for readings between zero and infinity, the needle coming to rest at the required position when the torques in the two coils are equal. As the coil system operates on the ratiometer principle, the deflection depends solely upon the value of the resistor under test: any variation in the generated voltage, due to varying handle speeds, affects both coils in the same proportion. The necessity of driving the tester at the correct speed is to develop sufficient voltage to find any possibly weak spots in the insulation under test.

When the instrument is to be used for continuity testing, a two-way switch is incorporated. In one direction of the switch, the instrument is set for insulation testing in the normal manner, whereas when the switch is set for continuity testing, the deflecting coil is connected in *parallel* with the resistance under test. Thus, maximum current flows through the deflecting coil when the terminals are open-circuited, and virtually no current flows through it when extremely high resistances are connected across the test terminals. Hence, the ratio of the currents in the two coils is opposite in direction to that which is obtained when the instrument is used for insulation testing, and for this reason the scale markings are also opposite in direction.

106. *Describe the construction and operation of a single-phase induction energy meter.*

The single-phase induction meter (Fig. 108) is essentially an a.c. instrument and comprises an upper and lower magnetic system between which is pivoted an

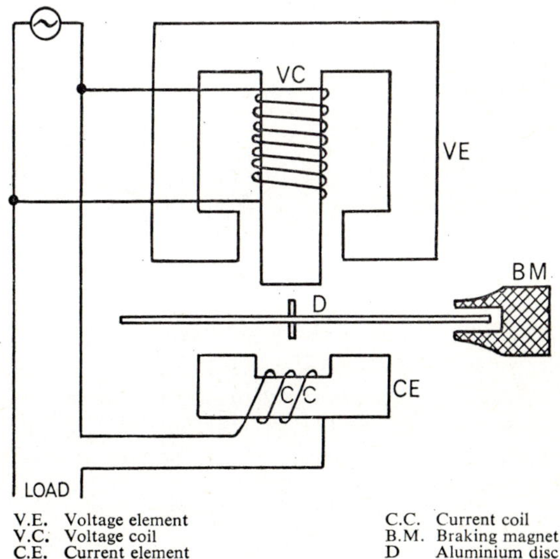

FIG. 108 *The single-phase, induction-type, energy meter.*

aluminium disc. In the upper magnetic system, which comprises the voltage element, a coil of many turns of very fine wire is wound on a laminated core. The coil, which is connected directly across the supply, has a high inductance, copper shading rings being fitted to increase the inductance, in order that the current may lag 90° behind the applied voltage.

The current element in the lower magnetic system consists of a coil wound on a laminated iron core, the coil consisting of only a few turns through which the current or part of the current in the circuit flows.

The aluminium disc rotates not only between the two magnetic systems but also in the field of a permanent magnet, which must be removed as far as possible from the electromagnetic fields, and whose purpose is to provide a braking torque. The aluminium disc is attached to a spindle which, in turn, is attached to a registering mechanism with gear trains and dials, although, in the modern form of energy meter, the dials are replaced by a digital system.

To understand the operation of the meter, reference should be made to Fig. 109 Because of the inductive nature of the voltage element, the flux Φ_v which it produces lags 90° behind the applied voltage V, whereas the flux Φ_i produced by the current elements is in phase with the current I, which lags behind the voltage by the phase angle ϕ, cos ϕ being the power factor of the circuit.

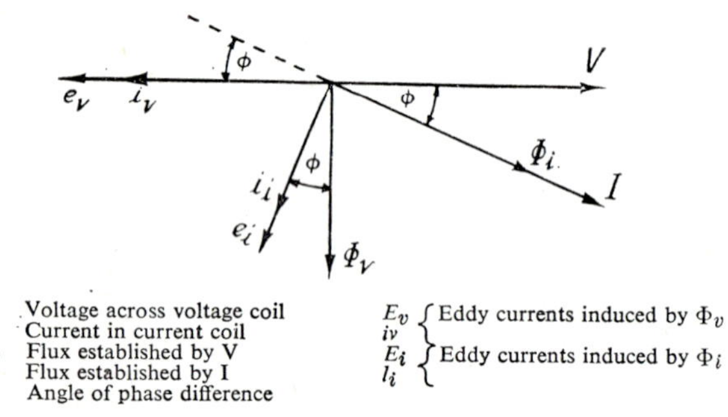

FIG. 109 *Vector diagram of an energy meter.*

Both Φ_v and Φ_i induce in the disc e.m.f.s e_v and e_i, both lagging behind the respective fluxes by 90°. These e.m.f.s, in turn, produce eddy currents i_v and i_i which are proportional to and in phase with them. Thus there is a phase difference of ϕ between i_i and Φ_v and one of $(180° - \phi)$ between i_v and Φ_i. The disc is therefore acted upon by two torques:

$T_1 = \Phi_v i_i \cos \phi$ which is proportion to $VI \cos \phi$.

$T_2 = \Phi_i i_v \cos (180° - \phi)$ which is proportional to $-VI \cos \phi$.

The resultant torque is thus proportional to the average value of the power.

The disc rotates against the eddy-current braking effect of the permanent magnet, uniform motion being established at a speed which is proportional to the power. Thus, the number of revolutions which the disc performs in a given time is a measure of the energy absorbed by the load during that period.

107. *What errors are introduced when using wattmeters for the measurement of power?*

Two methods are illustrated in Fig. 110 for the connection of wattmeters in single-phase circuits. In both cases, errors are introduced, the wattmeter indicating a value slightly higher than the actual power in the load.

With the connection shown in (a) the voltage coil is receiving its full voltage, but the current coil has a potential drop across it due to the current flowing through it. The wattmeter reading, in this case, is the power in the load plus the power loss in the current coil. With the connection shown in (b) the current in the current coil not only includes that in the load but also that in the voltage coil. Therefore, the wattmeter reading is the power in the load plus the power loss in the voltage coil.

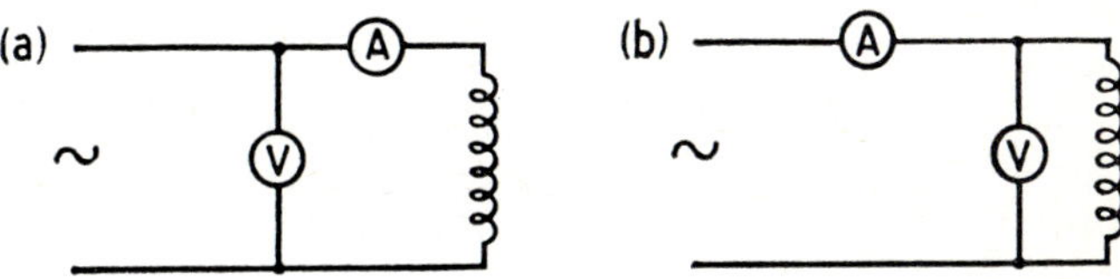

FIG. 110 *Illustrating the production of errors in wattmeter readings.*

The errors due to the losses in the instrument are normally small and can usually be neglected, but where a fair degree of accuracy is required, the method shown in (a) is preferable where the current in the circuit is low, while that shown in (b) is preferable where the current in the circuit is high.

Where extreme accuracy is required, it may be necessary to subtract the loss in the appropriate coil circuit from the wattmeter reading or, alternatively, to use a compensated wattmeter. This contains a current coil which is double wound, the extra compensating winding carrying a current in the reverse direction, so that the ampere-turns in the compensating coil neutralise the excess ampere-turns in the current coil due to the current carried by the voltage coil.

108. *Describe how the power in three-phase balanced and unbalanced systems may be measured.*

For three-phase balanced systems, only one wattmeter is required. The wattmeter, connected as shown in the star-connected system (Fig. 111), reads the power absorbed in the one phase. As the system is balanced, the total power is equal to three times the wattmeter reading. Where the power in a balanced delta-connected system has to be found and no neutral point is available, an artificial neutral point may be obtained, as shown in Fig. 112, by connecting three resistances of equal value across the three lines and connecting the voltage coil to the star point. Note that the voltage coil for standard supplies must be suitable for 240 V.

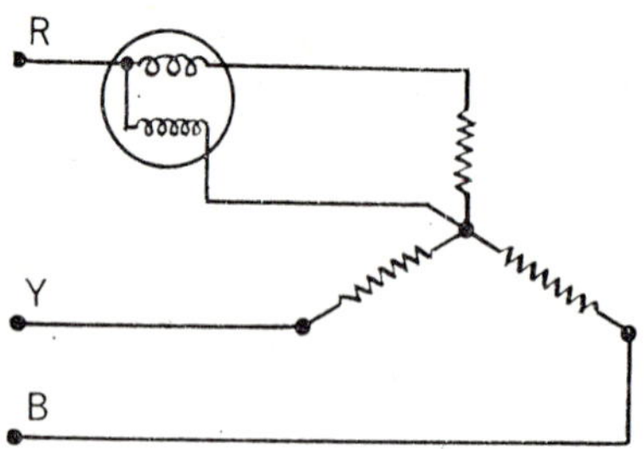

FIG. 111 *Measurement of the power in a balanced, three-phase, star-connected system with one wattmeter.*

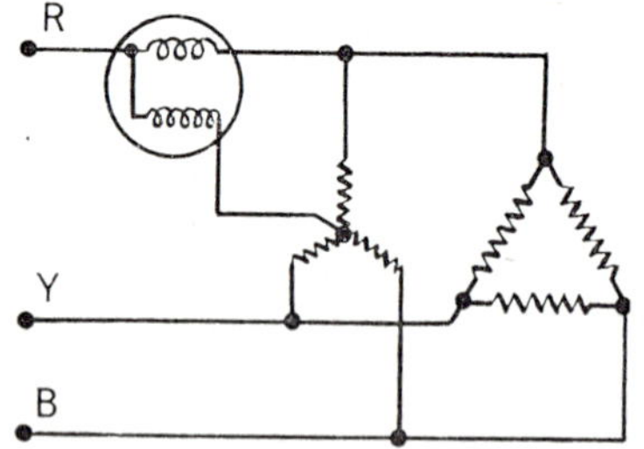

FIG. 112 *Measurement of the power in a delta-connected system using one wattmeter.*

For three-phase, four-wire unbalanced supplies, three single-phase wattmeters may be used, connected as shown in Fig. 113. As each wattmeter reads the power absorbed in each phase, the total power in the circuit is the sum of the wattmeter readings.

The two wattmeter method (Fig. 114) may be used to measure both the power and the power factor of three-phase, three-wire circuits for either balanced or unbalanced loads. It can be proved mathematically that if P_1 and P_2 are the wattmeter readings, the total power in the circuit is the sum of the two wattmeter readings, $P_1 + P_2$. It can also be proved that the power factor of the circuit can be found by using the formula:

$$\tan \phi = \frac{\sqrt{3}(P_2 - P_1)}{P_2 + P_1}$$

Once the tangent of the angle of phase difference has been found, it is comparatively simple to find the angle, then the cosine of the angle.

The cosine of the angle, however, may be found directly from the formula:

$$\cos \phi = \frac{1}{\sqrt{\left[1 + 3\left(\frac{P_2 - P_1}{P_2 + P_1}\right)^2\right]}}$$

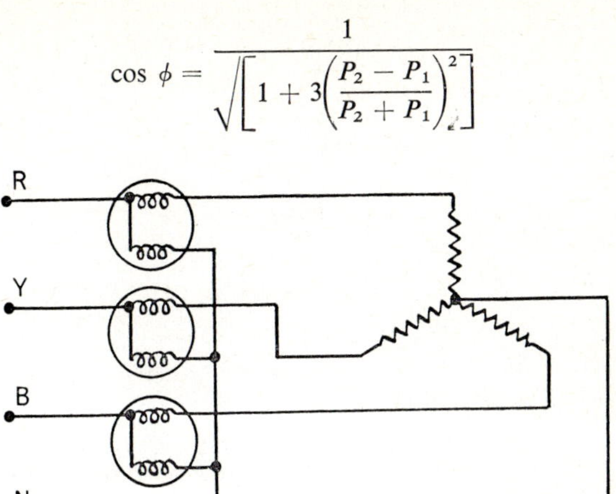

FIG. 113 *Measurement of the power in unbalanced, three-phase, star-connected circuits.*

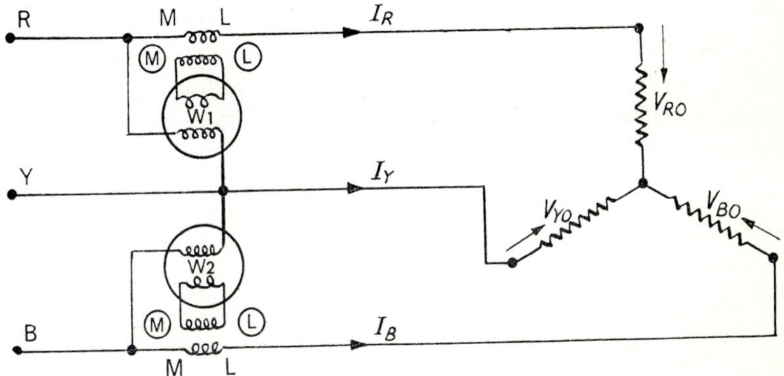

FIG. 114 *Measurement of the power by the two wattmeter method.*

12

D.C. motors

Elementary theory—commutation—armature reaction—speed control—Ward Leonard speed control—starting—fault finding.

109. *How is the movement of a current-carrying conductor in a magnetic field related to the direction of the current flow and magnetic field? Explain how this principle is used for reversing:* (*a*) *a shunt motor with interpoles,* (*b*) *a series motor, and* (*c*) *a compound-wound motor with interpoles.*

When the current is flowing in the direction shown in Fig. 115 (a) concentric lines of force are established round the conductor. These have the effect of strengthening the field above the conductor and weakening it below. Hence the conductor moves downwards (c). When the current is reversed (d) the concentric lines of force round the conductor flow in the reverse direction and therefore the field is now weakened above and strengthened below the conductor, hence it moves upwards. If the field is reversed and the current in the conductor remains in its original direction, a similar condition occurs as in (b) and the conductor again moves upwards.

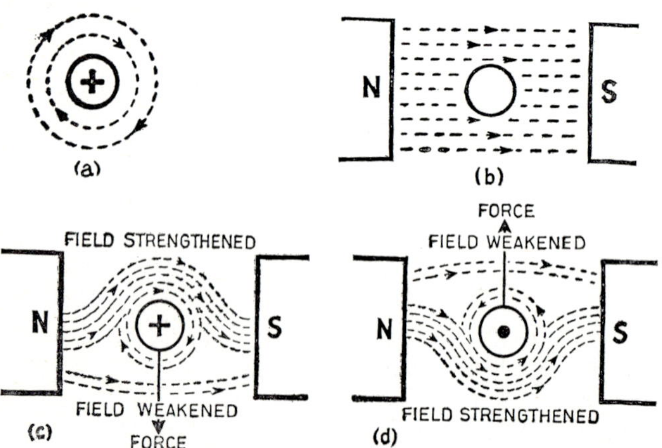

FIG. 115 *Force on a current-carrying conductor in a magnetic field.*

Fig. 116 illustrates how this principle is used for reversing a shunt motor with interpoles, either by reversing the field connections or the armature connections. Note that the interpoles must be reversed at the same time as the armature connections.

Fig. 117 illustrates the same principle when applied to the series motor, while Fig. 118 illustrates two methods of reversing a compound-wound motor. Note that if the field system is reversed, both the shunt and series windings must be reversed. Hence the simplest method is to reverse the armature winding.

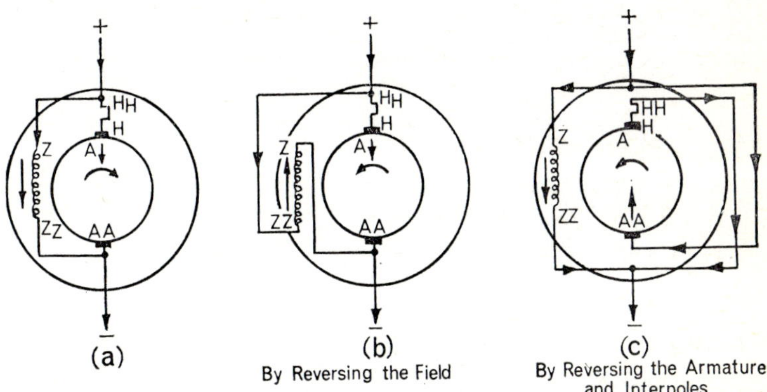

FIG. 116 *Reversing a d.c. shunt-wound motor with interpoles.*

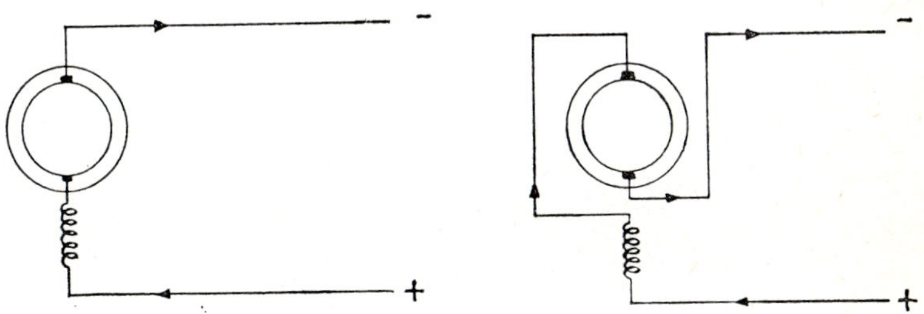

FIG. 117 *Reversing a d.c. series-wound motor.*

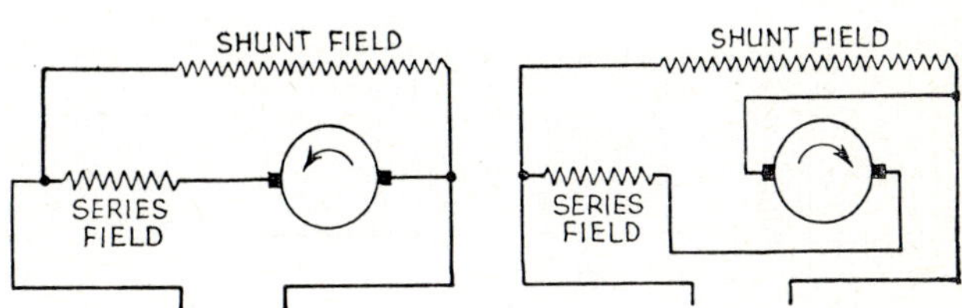

FIG. 118 *Reversing a d.c. compound-wound motor.*

110. *What are the essential differences between d.c. shunt, series and compound-wound motors, and for what type of drive is each most suitable?*

A certain strength of magnetic field may be obtained by having either a large number of turns carrying a low current or a small number of turns carrying a high current. Each type of magnetic circuit is used in shunt and series d.c. motors, and the compound-wound motor combines both types.

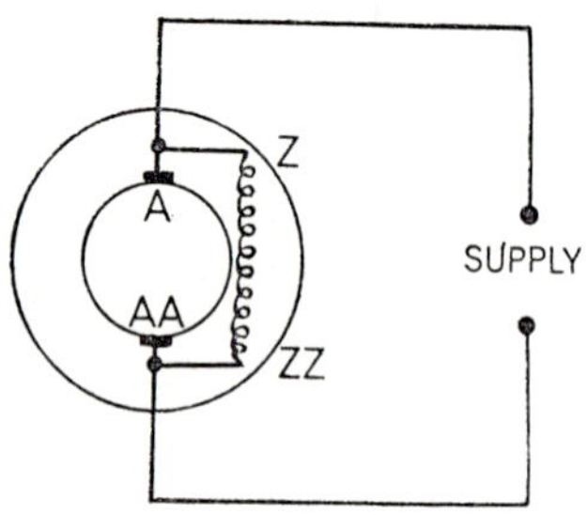

FIG. 119 (a) *D.C. shunt wound motor.*

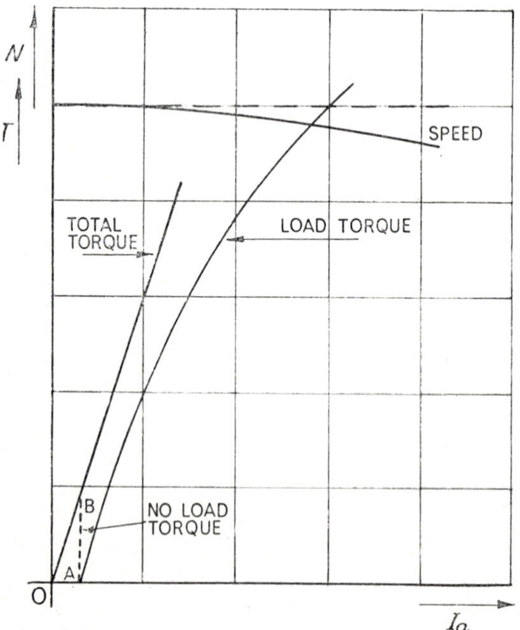

FIG. 119 (b) *Characteristics of a d.c. shunt-wound motor.*

In the shunt motor (Fig. 119 (a)), the field circuit is formed by winding a large number of turns of small diameter wires on iron cores, the windings being so arranged that alternate north and south poles are formed in the iron cores (referred to as the pole pieces). The whole winding is then connected in parallel with the armature circuit, so that the full supply voltage is impressed across the field circuit. The following expression:

$$N \propto \frac{V - I_a R_a}{\Phi}$$

relates the speed of the motor to the supply voltage, the voltage drop across the armature and the magnetic flux. Because the magnetic flux is constant and the voltage drop across the armature is only a small percentage of the supply voltage, the speed is virtually constant, dropping slightly as the current rises (Fig. 119 (b)). The torque is proportional to the magnetic flux and the armature current ($T \propto I_a\Phi$) and, therefore, as the flux is virtually constant, the total torque developed follows a linear law (Fig. 119 (b)).

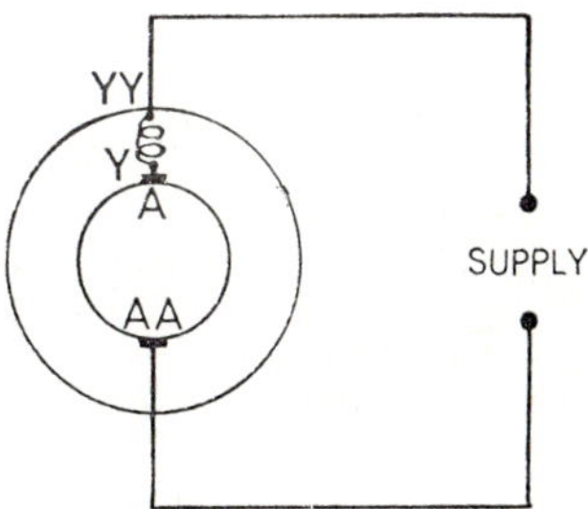

FIG. 120 *Series-wound motor.*

In the series motor (Fig. 120), the field circuit is formed by winding a small number of turns of large diameter wire on the pole pieces. The circuit is then wired in series with the armature across the supply and, therefore, the full load current flows through the field circuit. Consequently, at start, as the load current is high, a strong magnetic field is established, and, therefore, the speed is only low. As the motor accelerates, and the back e.m.f. starts to rise, the current falls, therefore. As the magnetic field also falls, the speed rises. If the fall in current is maintained, the magnetic field becomes so weak that the motor speed could rise to a dangerous level (Fig. 121). It is essential, therefore, that for this type of motor, the load never becomes disconnected. Also, because of the high current and strong field, a very strong torque is produced at start, but as the back-e.m.f. increases and the current falls, the torque is weakened. Where, however, the mechanical load increases, the current also increases, the torque being proportional to the square of the current. The torque curve follows that shown in Fig. 121. At high currents, the iron becomes saturated, and the curve flattens, the torque then being proportional to the current.

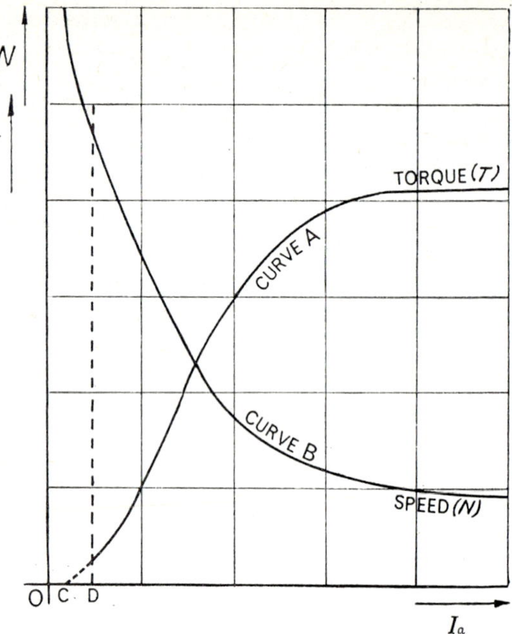

FIG. 121 *Characteristics of a series-wound motor.*

In the compound-wound motor both types of magnetic fields are used.

In the cumulative compound-wound motor, the field connections are so arranged that the series field assists the shunt winding, whereas in the differential compound-wound motor, the field is connected to oppose the shunt winding field. Hence the speed of the differential-type motor rises as the load increases, while the speed of the cumulative-type motor has a much larger drop than when the motor is run as a shunt motor only (Fig. 122). The series winding may also be connected as shown in Fig. 123 (a), in which case it is termed a long-shunt motor. If it is connected as shown in Fig. 123 (b), it is termed a short-shunt motor. In practice, it matters little whether the shunt is long- or short-connected. What is important is whether the series winding assists or opposes the shunt winding.

Shunt motors are used for drives where an approximately constant speed is required for all conditions of load, or where loads at various speeds of relatively long duration are required.

Series motors are used for drives requiring a large starting torque and slow starting speed, and where a constant speed is not required. Typical applications are found in electric traction and various types of hoists. Series motors should not be used where the load can fall to so low a value that dangerous speeds may be reached. For this reason, series motors should be coupled directly to the load and not through belt drives.

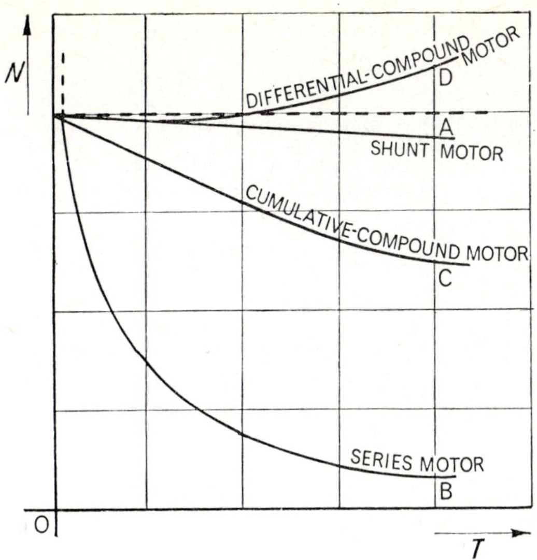

FIG. 122 *Characteristics of a compound-wound motor.*

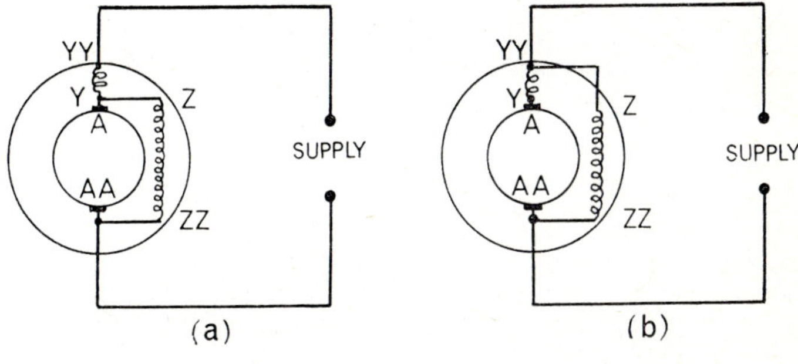

FIG. 123 (a)
Long-shunt compound-wound motor.

FIG. 123 (b)
Short-shunt compound-wound motor.

Because of its increasing speed characteristic with load, the differential compound-wound motor is seldom used, but the cumulative type is used for drives requiring a large starting torque and for fluctuating loads. The shunt winding prevents the speed from becoming excessive on light loads, and if a flywheel is fitted, a fall in speed through increased load enables the flywheel to surrender some of its kinetic energy so assisting the motor to cope with peak loads.

111. *How is the speed control of d.c. shunt and series motors most easily effected?*

The speed, the back e.m.f. and the magnetic flux of a d.c. motor are related as follows:

$$N \propto \frac{E}{\Phi}$$

which can be restated

$$N \propto \frac{V - I_a R_a}{\Phi}$$

From the above relation, it is clear that where the motor is connected to a constant voltage of supply, the speed of the motor can be varied in a downward direction by reducing the value of the back e.m.f., E. It can never be increased by this method as the back e.m.f. can never exceed the supply voltage. Although the voltage may be reduced by connecting a resistance in series with the motor, this method has little to recommend it as, although the voltage across the motor is reduced, this is partly off-set by the flux also being reduced which tends to increase the speed of the motor. As it is essential with the shunt motor that the flux remains constant, where speed control is obtained by varying the voltage across the armature, the simplest method is to connect a resistor in series with the armature, as shown in Fig. 124.

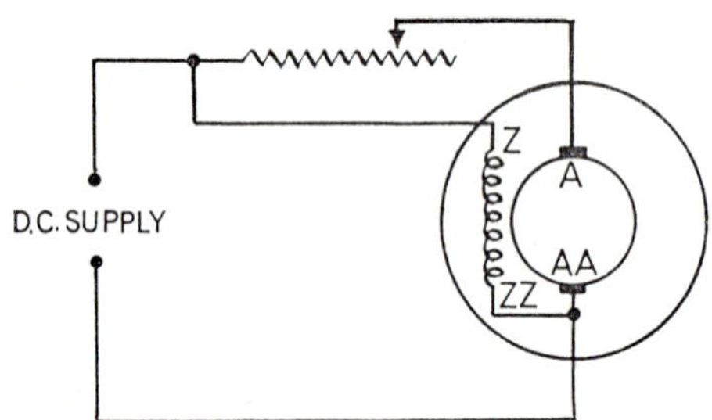

FIG. 124 *Reducing the speed of a shunt motor by armature resistance.*

The disadvantage of this method is that the resistor has to carry the armature current, and, therefore, large power losses may be incurred in the resistor. Where reduction in speed is necessary, Ward Leonard or electronic control is preferable especially for large machines.

In the series motor, the field circuit is in series with the armature and the connection of a resistance in series with both circuits must vary the voltage across the field and the armature. Therefore, the variation in speed has not the same relationship with variation of voltage as it has with the shunt motor.

As the supply voltage is normally constant, the only way of increasing the speed is by reducing the strength of the magnetic field. Thus, if the field strength is reduced by half, the speed is approximately doubled. The most effective way of reducing the magnetic flux is by connecting a variable resistor in series with the field circuit of a shunt motor (Fig. 125), or by the use of a diverter resistance in parallel with the field circuit in the series motor (Fig. 126).

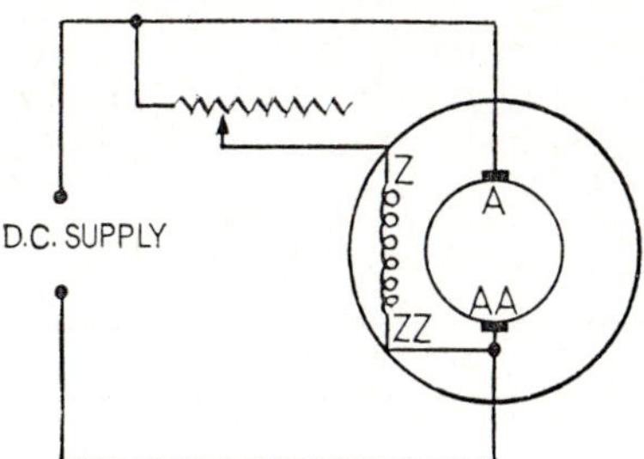

FIG. 125 *Increasing the speed of a shunt motor by field resistance.*

In the shunt motor, the field current is only a small proportion of the total current and regulation of speed is obtained with only a comparatively low loss of power. Therefore, the most economical method of controlling the speed is to design the motor so that a field controlling resistor can be incorporated.

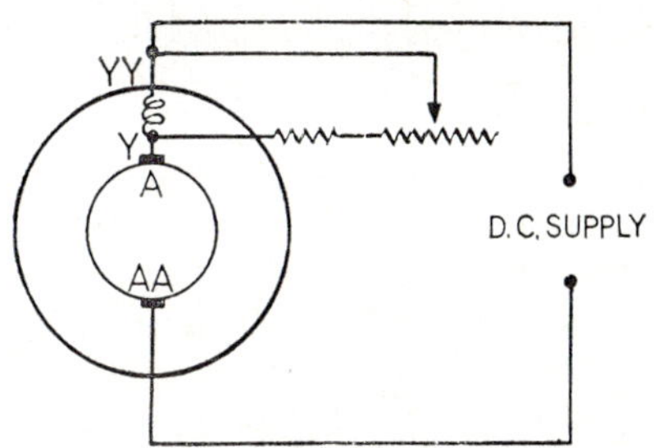

FIG. 126 *Increasing the speed of a series motor by diverter resistance.*

Connecting a resistor in series with the field in the series motor also affects the armature current, and, therefore, the magnetic flux variation is obtained by connecting a variable diverter resistor in parallel with the field circuit. Thus some of the current which would normally flow solely through the field circuit, flows through the diverter resistor.

If this is made variable then speed control may be effected. Maximum speed is obtained when the variable part of the resistor has been cut out, but it is essential that all the resistance cannot be cut out, otherwise the field circuit could be short-circuited and the motor speed attain a dangerous level.

112. *How may the speed of a d.c. shunt or compound-wound motor be varied over a wide range in both directions?*

A commonly used method for controlling the speed of d.c. shunt motors where a wide range of speeds is required is the Ward Leonard system (Fig. 127), in which the field excitation of the motor is constant, but the armature is connected to a variable voltage supply. The system is relatively expensive to install because, where the supply is alternating current, it normally necessitates the installation of four machines.

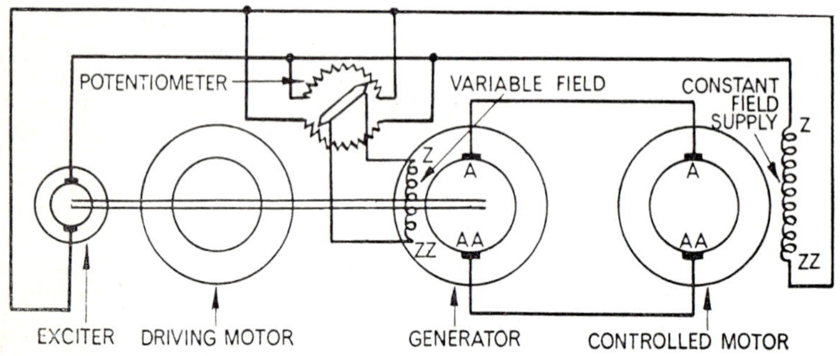

FIG. 127 *Variation of motor speed using a Ward-Leonard set.*

The variable-voltage supply for the armature of the motor is normally provided by a motor generator mechanically coupled to an a.c. driving motor driven from the a.c. supply. Also mechanically coupled to the a.c. motor is a small exciter used for supplying direct current to the field circuits of the work motor and generator; in many modern sets, the excite may be replaced by a rectifier.

The armature of the work motor is electrically connected to the armature of the generator, and since the speed of the a.c. driving motor is approximately constant, the output voltage of the generator is governed by its field excitation. Thus by incorporating a variable resistance in the field circuit of the generator, its output voltage may be varied over a very wide range. As this voltage is applied directly to the armature of the work motor, since the field of this motor is kept constant, its speed is proportional to the output voltage of the generator.

Thus there is very little needless expenditure of power in external resistances. By employing the potentiometer method of controlling the field excitation of the generator, the work motor may be driven in either direction with a smooth and wide variation of speed.

113. *Why is a series resistance necessary for the starting of a d.c. shunt motor, and explain, with the aid of a diagram, how the protective devices operate in such a starter?*

When the motor is operating and the armature conductors cut the magnetic field an e.m.f. is induced in them. This induced e.m.f. (back e.m.f.) opposes the applied e.m.f. so resulting in a corresponding reduction in the armature current. At start the conductors are at rest and there is no back e.m.f. induced in them. Therefore, if the motor was switched directly on to the supply, as their resistance is low, a high current would flow through them resulting in possible damage to the motor. It is therefore necessary to insert a resistance in series with the armature to limit the current to a safe value, normally to about one-and-a-half times full load current. The resistance is graded so that it can be cut out in successive stages as the motor accelerates and establishes its back e.m.f., until finally, all the resistance is cut out and the motor is running with its normal back e.m.f. to suit the corresponding load on the machine.

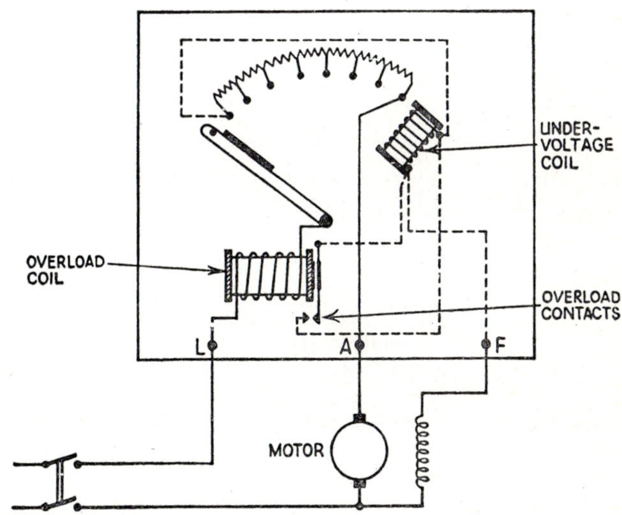

FIG. 128 *Starting a d.c. shunt motor.*

Undervoltage protection is provided by connecting a coil with a large number of turns of small diameter wire in series with the field circuit. When the handle is turned to the first stud, the starting resistance is connected in the armature circuit, the circuit to the undervoltage coil, and the field circuit made at the same time (Fig. 128). When all the resistance is cut out, a soft iron armature on the handle is attracted, and held in position, by the magnetic force established by the undervoltage coil. The handle remains in the 'on' position so long as the undervoltage coil is energised. Should the voltage fall below a pre-determined level, the undervoltage coil is de-energised sufficiently for the spring attached to the handle to return the latter to the 'off' position.

The excess-current device comprises an electromagnet connected in series with the armature. As the armature resistance is low, the resistance of the device must be considerably lower so that little voltage is lost across it. It is, therefore, wound of a small number of turns of relatively large cross-section. The device is made adjustable so that it can be arranged to operate when the current reaches a pre-determined level. Attached to the armature is a fixed contact which, when the armature moves inward, bridges two contacts connected respectively to each end of the undervoltage coil. The bridging of these two contacts short circuits the undervoltage coil with a low resistance path, so that the coil is de-energised, and the handle is returned to the 'off' position.

This method is preferable to opening the field circuit, because, if the latter method were adopted, the dissipation of the energy stored in the magnetic field would create a severe spark across the overload terminals. With the former method, the energy is safely dissipated in the armature circuit.

It is essential that the undervoltage coil is connected to the first terminal of the resistor and not the last, since in the latter case, the potential across the field circuit would be the same as that across the armature circuit, and there would be a serious reduction in torque with a possibility of the motor failing to start.

114. *What are the main faults that occur in d.c. machines and how may they be detected?*

The major faults that occur are (1) breakdown of the insulation on the brush holders; (2) open- or short-circuits in parts of the field system, and (3) open- or short-circuits on the armature, which also includes faults on the commutator.

(1) This fault should seldom occur if the machine is regularly serviced, but can be usually detected by the appearance of charring of the insulation and tracking when the machine is in operation.

(2) These faults may be detected by checking the resistance values of each coil, either by means of a Bridge Megger or by using an ammeter and voltmeter. A short circuit can often be detected by touch, the sound coils feeling hotter than the faulty coil when the machine has been running for an appreciable period of time. An open circuit is indicated by a zero reading on the ammeter and an infinity reading on the Bridge Megger.

(3) Although blackening of the commutator and burning and pitting of the commutator segments will often indicate that an armature is faulty, this may be proved by means of a drop test, as follows:

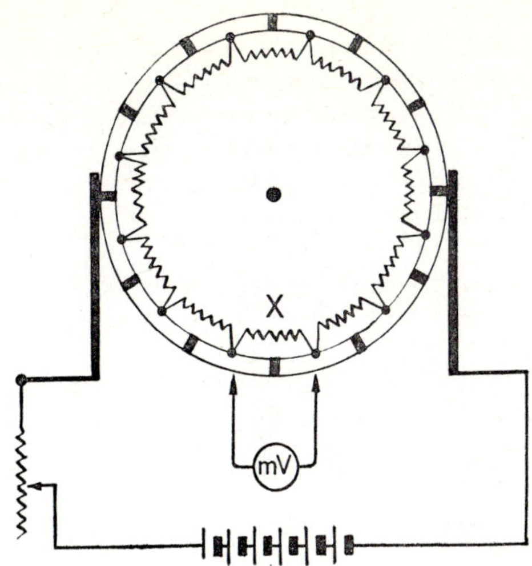

FIG. 129 (a) *Drop test on a d.c. armature.*

As all the armature coils have approximately the same resistance, if a source of low potential is connected across diametrically-opposite commutator segments (Fig. 129 (a)), each of the armature coils in each path should have the same potential drop across it, if the armature is in a serviceable condition. Should a fault occur on one coil, the potential drop across it will vary, the nature of the fault being indicated by careful consideration of the readings taken. Thus, a short-circuit on a coil at X will give a zero reading on the millivoltmeter when it is connected across the coil. On a sound armature the reading of the millivoltmeter across each coil should be approximately the same.

In testing for open-circuits, the meter should be capable of reading the full voltage of the battery. If an open-circuit occurs, no potential difference will appear across the sound coils in that path, whereas, full potential will appear when the voltmeter is connected across the break.

In testing for short-circuits, a low reading may well indicate a partial short-circuit. This may be due to carbon dust, or fragments of copper between segments on the commutator. In this case, if the insulation is carefully cleaned, the readings should return to normal. If, however, the reading still remains low, the fault is still on the armature coil winding.

Earth faults may also be detected by means of the drop test, one of the supply leads being connected to the armature shaft and the other to the commutator. The millivoltmeter is connected between the shaft and the individual segments respectively. If different readings are obtained, the lowest millivoltmeter reading will indicate the point nearest to the earth fault.

Before putting the armature back into service, any grooves or flats on the commutator should be removed by skimming the surface in a lathe. Where skimming is resorted to, care must be taken to ensure that although the surface is made perfectly true, too much metal is not removed from it. After skimming, and in many instances where skimming has not been resorted to, the micas must be undercut. Fig. 129(b) illustrates correct and incorrect undercutting of the mica insulation. In carrying out this work, only a correct undercutting tool should be used, and extreme caution should be employed to ensure that scoring of the copper segments is prevented.

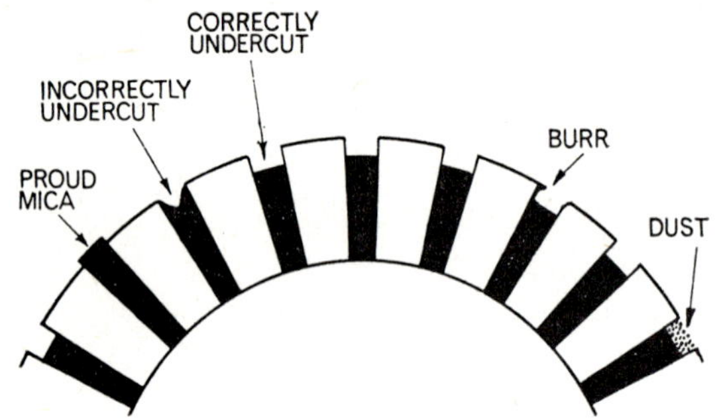

FIG. 129 (b) *Correct and incorrect undercutting of a commutator.*

Also before putting it back into service, the complete motor should have an insulation test carried out on it, using an insulation tester with an output of not less than 500 V. The reading obtained then should be:

$$\frac{\text{rated volts}}{1000 + \text{rated output power}}$$

for large machines and not less than 1 MΩ for fractional horsepower machines.

115. *Tabulate the faults and the checks that should be made to locate them on d.c. machines.*

See Table 3 (pg. 199).

Table 3.

Fault	*Possible cause*	*Remedy*
Motor fails to start	(1) Failure of supply (2) Stalling due to overloading (3) Friction (4) Brushes away from commutator (5) Open field-circuit (6) Brushes in wrong position (7) Open circuit in armature winding (8) Short-circuit in field winding	Check control gear, etc. Reduced load Check for mechanical defect Ease brushes in holders and adjust spring tensions if necessary Test for continuity Check brush position Test by drop test or Growler; rewind, if necessary Check resistance of each coil
Motor rotates in wrong direction	Connections to field or armature reversed	Reverse connections to either field or armature
Poor torque	(1) Dirty commutator (2) High resistance in field circuit (3) Friction	Clean commutator Test for continuity Check for mechanical defect
Reduced speed	(1) Overloading (2) Defective armature coil (3) Friction	Reduce load Test and rewind if necessary See above
Overspeed of shunt or compound motor	Weak field	Check field connection and insulation
Overheating	(1) Overloading (2) Low insulation resistance (3) Partial shorts in field system	Reduce load Put on insulation test Dry out, if necessary Check resistance of each coil. Rewind, if necessary
Uniform sparking at all brushes	(1) Dirty commutator (2) Excessive load (3) Incorrect brush position (4) "Proud" micas	Clean commutator Reduce load Check brush position Undercut micas
Continuous sparking at one brush	(1) Brush sticking (2) Brush not bedded (3) Insufficient spring pressure (4) Brush too short (5) Dirty brush	Free brush Bed Brush Test with balance and adjust Fit new brush Clean brush contact
Intermittent sparking at all brushes	(1) Flats on commutator surface (2) Individual "proud" micas (3) Grooving of commutator (4) Open circuit in armature winding (5) Short circuit in armature coil (6) Segments short-circuited	True the commutator Undercut micas Skim commutator Test by drop test or Growler; rewind if necessary As above Clean out commutator slots, then test

13

A.C. motors

Production of torque in three-phase motors—protective devices—starting squirrel-cage motors—starting wound—rotor motors—speed control of three-phase motors—single-phase—reversing—single-phase motors—split-phase motor—capacitor start and capacitor start-and-run motors.

116. *What factors affect the choice of motor for selected drives?*

The following factors affect the choice of motor:

(1) What horsepower is the motor required to transmit?
(2) What is the speed required to suit a particular drive?
(3) What are the electrical supply conditions?
(4) What are the starting conditions?
(5) What are the atmospheric conditions?
(6) What are the load conditions?
(7) What are the mechanical conditions?

(1) The motor must be capable of withstanding sudden surges of energy and yet operate at an optimum loading. Where induction motors run lightly loaded, the power factor of the circuit may be very low, which, if repeated, could lead to the installation of expensive power factor correction equipment

(2) The speed of the normal induction motor is determined by the frequency of the supply and the number of poles it contains. Higher speeds may be obtained by installing special machines, but for relatively low-speed drives it may prove more economical to install high-speed motors with reduction gear than low-speed induction motors coupled directly to the load.

(3) Here, it is necessary to know what type of system is available. Is there any appreciable voltage variation? Is there any limit to the maximum load to be connected? Is current limitation necessary at starting? What types of tariff are available?

(4) Here, it is necessary to know the frequency of starting, the number of reversals (if any), and the method of starting. Does the inertia of the load entail long running-up periods (with direct-on-line starting it is necessary that the motor attains its maximum speed almost instantaneously)? Is hand-operated or automatic starting required with or without remote control?

(5) The motor must be selected to suit the conditions in which it is to be installed and the following gives some idea of the type of motor suitable for the required situation.

Protected and screen-protected motors. These may be installed in all situations where the atmospheric conditions are suitable, that is where dust and dirt are not very prevalent. They are suitable for floor and wall mounting but not for vertical mounting unless protection is given against falling dirt and moisture.

Drip-proof motors. This motor is similar to the protected type but is fitted with a cowl to exclude falling water, moisture, dirt, etc.

Totally enclosed (fan-cooled) motor. While not all totally enclosed motors are fitted with fans, the most popular and important type is the totally enclosed fan-cooled machine, which incorporates a mechanically protected fan for cooling purposes. The surface is usually ribbed to increase the heat dissipation.

This type of motor is weatherproof and resists the action of water, but is not waterproof. It is therefore suitable for situations where excessive dust or moisture is present, but not in positions where it is liable to be submerged.

Dust-proof motor. This type of motor has now generally superseded the pipe-ventilated type, both types being recommended where the working conditions are extremely onerous, or where the cooling fan is likely to cause atmospheric disturbance, particular applications being flour mills, cement works, paper mills and paint and varnish works, etc.

Flameproof motors. A flameproof motor is one which will withstand without injury any explosion of prescribed gas that may occur within, under practical conditions of operation within the rating of the apparatus (and recognised overload), and will prevent the transmission of flame such as will ignite any prescribed inflammable gas which may be present in the surrounding atmosphere.

This calls for a motor constructed with specially machined faces which virtually prevent the entry of any flammable gas. To obtain a 'Buxton' certificate from Her Majesty's Inspector of Mines, the motor is charged and surrounded by flammable gas. The internal charge is ignited and the windings must suffer no damage, nor must the surrounding gases be ignited. The applications of such motors are in situations where explosive or flammable gases or dusts are to be found, such as in coal mines, fuel plants, petroleum-pumping stations, gas plants, chemical works, etc.

Where flameproof motors have to be installed, it is also necessary for all the associated equipment to be of the flameproof type, unless it is installed outside the prescribed danger area.

Oil-filled sleeve bearing motor. Where precautions have to be taken against noise, the oil-filled sleeve bearing motor is generally preferred to the ball or roller bearing motor. Instead of the shaft being supported in ball races, it is supported on a cushion of oil, the oil being supplied from a reservoir beneath the bearing with oil rings circling round the shaft keeping the latter well lubricated. In addition, the motor and the driven machine should be mounted on an anti-vibration bed of cork or other suitable anti-vibratory material. The control gear should also be designed for silent operation, and to assist in eliminating magnetic hum, a rectified d.c. contactor coil may be used.

(6) It is clear that where a motor is continuously loaded, the temperature rise must be greater than in a motor where the same magnitude of load is applied for only short periods of time; for this purpose, B.S. 2613 defines two classes of ratings. The full-time rating defines the load for which the motor may be run for an unlimited period, while the short-time rating defines the load at which the motor may be operated for the period and conditions specified on the rating plate at a specified ambient temperature and altitude. The two short-time ratings are for one hour and half an hour periods.

(7) When considering the mechanical conditions, it is necessary to remember that, when motors are being mounted, ball or roller bearings are suitable for horizontal or vertical mountings, but sleeve bearings must not be used for vertical mountings.

Motors may be damaged through vibrations transmitted from the driven machine, and where such vibrations exist, the root cause must be investigated and eliminated.

Where end-thrust is likely to occur in a motor, the bearings must be of the thrust type. Where large-diameter pulleys or heavy belt loads are being used, it is advisable to reduce the load on the driving-end bearing of the motor by using a third bearing.

Also when considering mechanical conditions, the type of drive to be used is important. Thus flat belt drives are normally satisfactory where the distance between pulley centres exceeds three times the larger pulley diameter, and the ratio of pulley sizes is not greater than 4:1. Where flat belts are used, the drive should be arranged with the slack side uppermost. Slide rails should be used, enabling the motor to be lined up and ensuring correct belt tensioning. Belts should not be too tight, nor should belt fasteners liable to knock the pulley be used. The maximum belt speed should not normally exceed 1300 m/min, and, generally speaking, vertical drives should not be used. Where fast and loose pulleys are used, to avoid damage to the bearing, the drive to the fast pulley should be taken from the end of the pulley nearest to the motor bearing.

Vee-belt drives may be used satisfactorily for any type of drive but are especially preferred to flat belts where the distance between the driver and driven pulleys is short. Where more than one belt is employed, the belts should be of equal length, although a slight variation is permissible. Enough belts should be chosen so as to give an additional margin of power and also to reduce belt slip. A belt speed of up to 1300 m/min with an arc contact on the pulley of 180° is normally satisfactory, but in certain cases belt speeds up to 2000 m/min are permissible, but the arc contact must not drop below 120°.

Where space does not even permit the fitting of Vee-drives, it is often advantageous to fit a chain drive. It is possible to arrange for much higher speed ratio with this type of drive (in the order of 8:1) than with other types of drive, so enabling high-speed motors to be employed. When installing chain drives it is essential to ensure correct alignment between the driver and driven sprocket wheels, and the driver sprocket should be as close as possible to the bearing cap.

Direct drives are employed where the driven machine can be driven at the same speed as that of the motor, such as for motor-generators, fans, pumps and other similar drives. To compensate for small inaccuracies in mounting, a flexible coupling should be incorporated in the drive, and care should be taken to ensure that there is a gap between the two ends of the shafts inside the coupling. Where solid couplings have to be used, correct alignment is essential to prevent the possibility of a broken shaft. A small amount of end play is also essential so that, when set, the rotor may move approximately 1·5 mm in either direction.

117. *Describe the usual methods of starting three-phase induction motors, and state under what circumstances each method would be used.*

Most small three-phase squirrel-cage motors, and sometimes large ones, are started by using a direct-on-line starter. In this method, the motor is connected directly to the supply through the starter without employing any current-limitation device. The normal three-phase squirrel-cage motor takes approximately six times full-load current when started direct-on-line with a starting torque approximately 1·5 times full-load torque. As most installations, even small ones, are capable of absorbing the relatively high starting current, and as the majority of drives require a torque well within the range of this type of starting, this method is almost universally used for motors up to 3·75 kW. In large industrial installations, where the high starting current can easily be absorbed, much larger motors are started direct-on-line.

For motors exceeding approximately 3·75 kW in small installations and for higher rated motors in large installations, some form of current limitation is necessary at starting. One widely adopted method of starting cage-type three-phase motors is the star-delta method. With this type of starting, the motor windings are connected in a star formation to the supply at start and in a delta formation in the running position. In star, this has the effect of reducing the voltage across each winding by $1/\sqrt{3}$ and the current in the lines by $\frac{1}{3}$. Hence, at start, the line current is approximately twice the full-load current. Unfortunately, the starting torque is reduced to approximately $\frac{1}{2}$ full-load torque. Hence, this type of starting is only suitable for drives where the inertia of the load at start is so low that a motor with a low starting torque is suitable.

Another method of starting cage-type motors, where current limitation at starting is required, is by employing an auto-transformer starter. The principle is somewhat similar to the star-delta method, in that the voltage across the windings is reduced during the starting period. In this case, however, the reduction in voltage is achieved by connecting the windings across the secondary winding of an auto-transformer, but, as the secondary winding contains three tappings, the reduction can be varied by means of adjusting the tapping on the secondary winding by a special three-way switch. Thus the tappings may be arranged to give possibly 40 per cent, 60 per cent or 80 per cent of the voltage across the windings.

The line current and the torque, however, are reduced in proportion to the square of the reduced voltage across the windings. The 60 per cent tapping gives similar values to that obtained by the star-delta method, whereas the higher tapping give a higher starting current and torque and the lower tapping a lower starting current and torque. Thus, if only a very low starting torque is required, the lowest tapping may be used. If a torque similar to that given by star-delta starting is required, the second tapping may be used, and if a torque approximately equal to the full-load torque of the motor, the highest tapping may be used. Although more expensive to install than the star-delta starter, the greater range of starting torque makes the starter more suitable where the inertia of the load at start varies. It must also be remembered that the starting current varies in proportion to the starting torque and can be quite high when the highest tapping is used.

Where current limitation is necessary at starting and yet a good starting torque is required, both the star-delta and auto-transformer methods of starting are unacceptable, and therefore a wound-rotor motor would be preferred to the squirrel-cage motor. In the squirrel-cage motor, the conductors are solid and are welded together at both ends of the rotor so as to form a closed circuit, whereas in the wound-rotor motor, the conductors are wound on the rotor, one end of each windings being connected together to form a star point, the other three ends being taken to slip rings mounted on the shaft. This enables connections to be taken to three variable star-connected resistors. By inserting resistance in the rotor circuit, the impedance of the circuit is increased, so reducing the total current in both the rotor and stator circuits. The introduction of resistance in the circuit, however, improves the power factor and also increases the current in phase with the voltage (the current which is responsible for the production of torque), so that an increase of torque is produced at the same time as the total current is reduced. This motor is therefore installed where current limitation is required together with a good starting torque.

118. *Describe the essential items of the control apparatus for electric motors and state for what purpose they are included.*

Every electric motor, subject to certain exemptions, must be provided with undervoltage protection to prevent automatic re-starting in the event of a drop in voltage or failure of the supply. Undervoltage protection is provided to ensure the safety of personnel should the supply fail or fall below a pre-determined level. For instance, if undervoltage protection is not provided, and the supply fails, there is a grave risk to personnel should the supply be restored without warning to an operative who is perhaps cleaning or working on a machine tool. If the isolator has not been switched off, the machine would instantly start up when the supply was restored. With undervoltage protection, some manual operation is necessary before the motor can be started (that is, pressing a push button to operate the starter.)

In certain cases, the undervoltage protective device may be omitted where irregular starting from automatic control gear (thermostats, float switches etc.,) is required, or where a dangerous condition might occur should the motor fail to re-start after a temporary interruption of the supply.

Means of isolation must also be provided, suitably placed, so as to disconnect the motor and all its associated apparatus from the supply. The means of isolation is used not only for disconnecting the motor and control gear for maintenance purposes, but also to disconnect the motor should the starter fail to operate when so required, hence the necessity of installing the starter and means of isolation adjacent to the motor.

One means of isolation, however, may be installed to control a group of motors, provided that the isolation of all the motors at the same time is acceptable.

Where the motor exceeds 0·375 kW, a suitable device must be incorporated to protect the motor and the cables between the motor and the starter against excess currents. In all push-button operated a.c. starters, a coil is necessary for undervoltage protection, and by including a device with a suitable switch in the coil circuit, the same coil may also be used to give excess-current protection. The device is usually calibrated so as to give protection against varying degrees of overload, and where undervoltage protection is not provided, the excess-current device should be of the re-set pattern.

119. *Describe two types of excess-current device normally associated with a.c. motor starters and state the reasons why time-lags are an essential part of the device.*

The two types of excess-current protective device associated with a.c. switchgear are:

(1) the magnetic type,

(2) the thermal type.

The magnetic type of overload device (Fig. 130) has an electromagnet connected in each line wire between the supply and the motor. The coil encloses a soft-iron plunger mounted on a spindle, to which is also attached a disc immersed in a reservoir of oil. The inertia of the oil imposes a braking action on the disk, retarding the movement of the soft-iron plunger in the electromagnet. When the normal operating current of the motor is flowing through the overload coil, an attractive force is exerted on the plunger but is not strong enough to overcome the inertia of the oil. If excess currents of high magnitude but of

FIG. 130 *Electromagnetic overload device.*

short duration flow through the overload coil, the attractive force on the plunger is very strong, but because of the short time duration is still unable to overcome the inertia of the oil. If the overload is sustained for a sufficient period of time, the attractive force overcomes the inertia of the oil, and the plunger moves upward, either mechanically tripping the starter or causing two contacts in the coil circuit to open, de-energising the coil and tripping the starter. The container of oil is screwed on to a boss, so enabling the device to be set to operate at the prescribed overload, and, it is essential therefore, that only the correct grade of oil should be used and that the container should be filled to the right level.

The thermal overload device (Fig. 131) has a heating coil in each line, connected between the supply and the motor terminals, and a bimetallic strip, the movement of which causes the tripping device to operate. The normal operating current is not sufficient to cause the strip to move far enough to open the tripping contacts, but when the overload is sustained, the temperature of the heating coil is raised sufficiently to cause the bimetallic strip to bend against a bar mechanically connected to the trip contacts; the movement of the bar eventually causes the contacts to be opened. The time delay is created by the time taken for the bimetallic strip to move the necessary distance to open the overload contacts.

Time lags are used so that the starter will not trip on transient overloads, but only when the overload is sustained and of sufficient magnitude.

FIG. 131 *Thermal overload device.*

120. *Describe how to construct a schematic diagram for a three-phase, direct-on-line starter. From the schematic diagram construct the full starter diagram and explain its operation. Draw a further diagram to show how the starter can be converted to operate from a float switch or thermostat.*

Fig. 132 illustrates how the schematic diagram is progressively constructed: (a) shows how the contactor coil is energised when the start button is pressed, and also how it is de-energised when the finger is removed; (b) shows how the coil remains energised by incorporating a retaining contact, made simultaneously with the main contacts. With (b) alone, the motor could only be stopped by breaking the main circuit to the starter, for example, by opening the main switch controlling the entire circuit, therefore a stop button is included in the circuit (c), which is normally closed but which, when pressed, breaks the coil circuit. (d) shows how, when the overload contacts are opened through excess current flowing in the main circuit, the coil circuit is again broken, tripping the starter. (e) illustrates the complete schematic diagram including a remote stop-and-start push button position. It also illustrates one fundamental principle concerning stop and start buttons, that is all start, buttons are connected in parallel with one another and all stop buttons and other stopping devices are connected in series with one another, a link being incorporated in the starter for this purpose.

To construct the starter diagram from the schematic diagram, first the three main line wires are drawn from the underside of the L_1, L_2 and L_3 contacts to the A, B and C terminals, respectively. Starting at the top side of the L_1

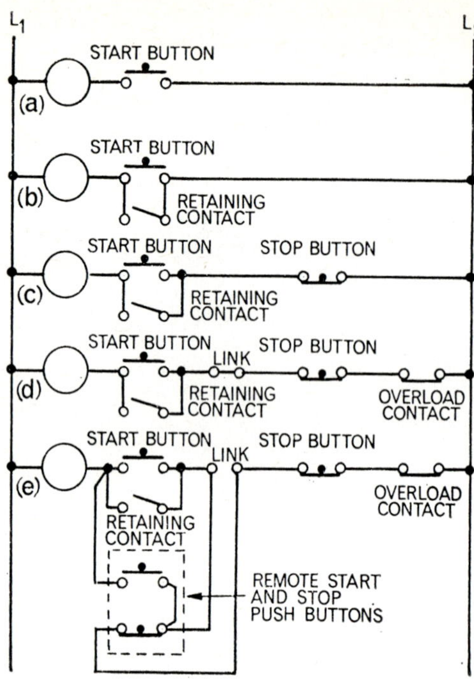

FIG. 132 *Schematic diagram for a direct-on-line starter.*

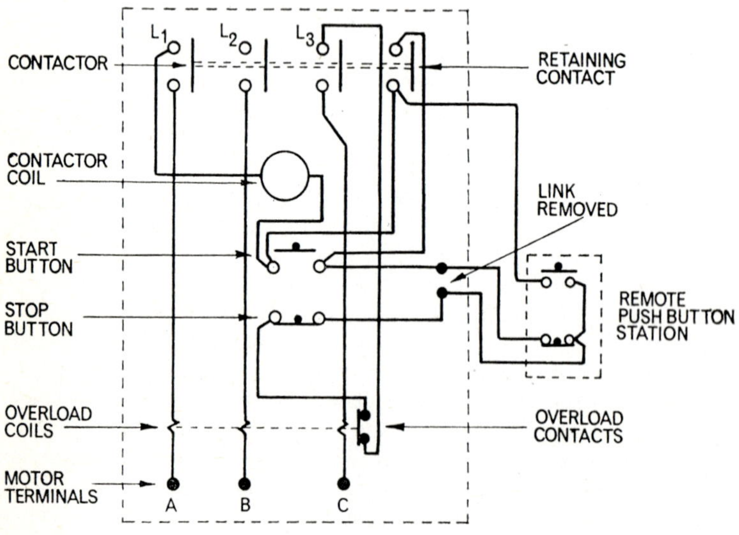

FIG. 133 *Direct-on-line three-phase starter.*

contact, the control circuit is drawn in the sequence shown in the schematic diagram Fig. 132, finishing at the top-side contact of either L_2 or L_3. The completed diagram should then be similar to the one illustrated in Fig. 133.

When either the start button in the starter or the remote push-button station is pressed, the contactor coil is energised, closing the main contactor. This completes the circuit from the three lines, through the overload coils, to the motor. At the same time the retaining contact is also closed, so keeping the contactor coil energised once the start button has been released. The motor will then continue to run until the stop button is pressed or one of the other stopping devices operates.

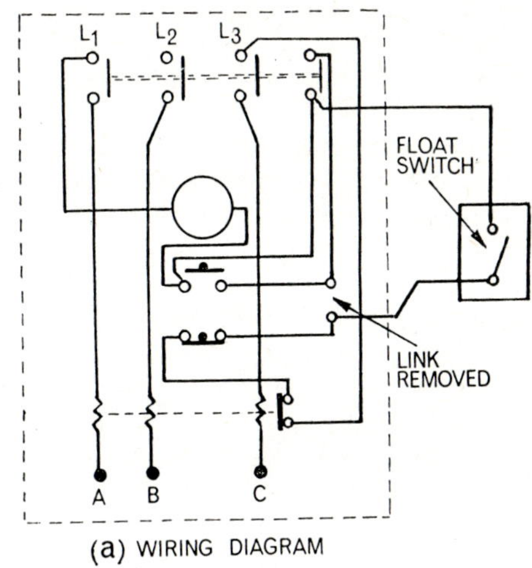

(a) WIRING DIAGRAM

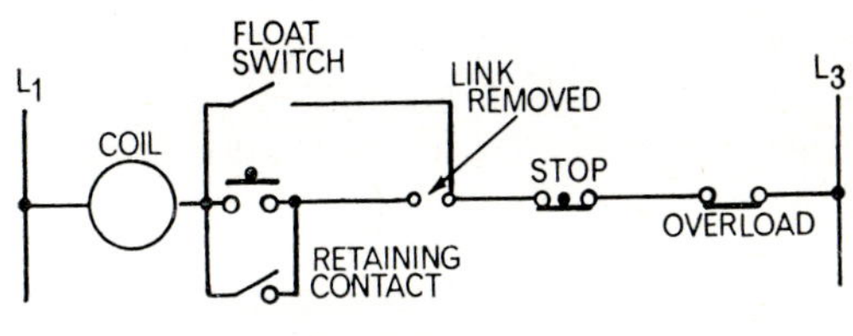

(b) SCHEMATIC DIAGRAM

FIG. 134 *Illustrating two-wire control.*

In many circumstances, the starter has to be automatically controlled by means of a device such as a float switch or thermostat. Fig. 134 shows how the starter circuit of Fig. 133 is converted to operate from a two-wire control circuit. For this type of control, the retaining contact and the stop-and-start buttons serve no useful purpose and could be discarded. It is also clear that if the overload device is not of the type which, when once operated, requires some form of manual operation to be brought into service again, the starter will be continually opening and closing should a sustained overload occur. For this type of circuit then, the overload device should be of the re-set pattern. Should a dangerous condition be likely to occur through the overload device tripping the starter and remaining unnoticed, some form of alarm should also be installed.

121. *Describe, with the aid of a diagram, the operation of a hand-operated star-delta starter and explain why it is essential to pause before changing over from star to delta. Is it necessary for the cables between the motor and the starter to be of the same size as the sub-circuit cables?*

The hand-operated star-delta starter (Fig. 135) incorporates a change-over switch which, in the start position, connects the motor in a star formation to the supply and, in the run position, connects it in a delta formation. Undervoltage protection is provided by means of an electromagnet, which is energised when the handle is in the run position, a catch holding the handle in the run position unless the voltage falls below a pre-determined value. Excess-current protection is also provided for, in this instance, by means of an overload device of the electromagnetic dash-pot type, which operates the same tripping mechanism as the undervoltage device. The tripping mechanism may also be operated manually.

Six conductors are taken from the starter to the motor and great care should be taken to ensure that the cables are connected to terminals of the motor with the corresponding letters of the terminals in the starter.

When the handle is in the start position, the top six switches are closed. This connects the terminal A_2, B_2 and C_2 through three of the switches to the respective three line terminals, while the terminals A_1, B_1 and C_1 are connected in a star formation through the other three switches. Neither the undervoltage solenoid, or the overload coils are connected in the start position.

When the motor has accelerated sufficiently, the handle is changed to the run position. This operation must be performed smartly, otherwise a catch prevents the handle from reaching the run position. This same catch also prevents the motor from being started in the delta position.

In the run position, the switch is changed over, connecting the six centre terminals to the six lower terminals. This connects A_2 to C_1, B_2 to A_1 and C_2 to B_1, each pair of terminals being connected to one of the supply lines. In this position, the undervoltage and excess-current devices are now incorporated in the circuit.

The reason why the excess-current device is not included in the start position is that, if the motor is substantially loaded, a relatively long period of time will elapse before the current falls to a value approaching its full load value, in which case the excess-current device may be operated. Hence the device is only included when the motor is running in the delta position.

The necessity for pausing before changing over the switch to the run position is that, if the motor has not accelerated sufficiently for the back e.m.f. to reach

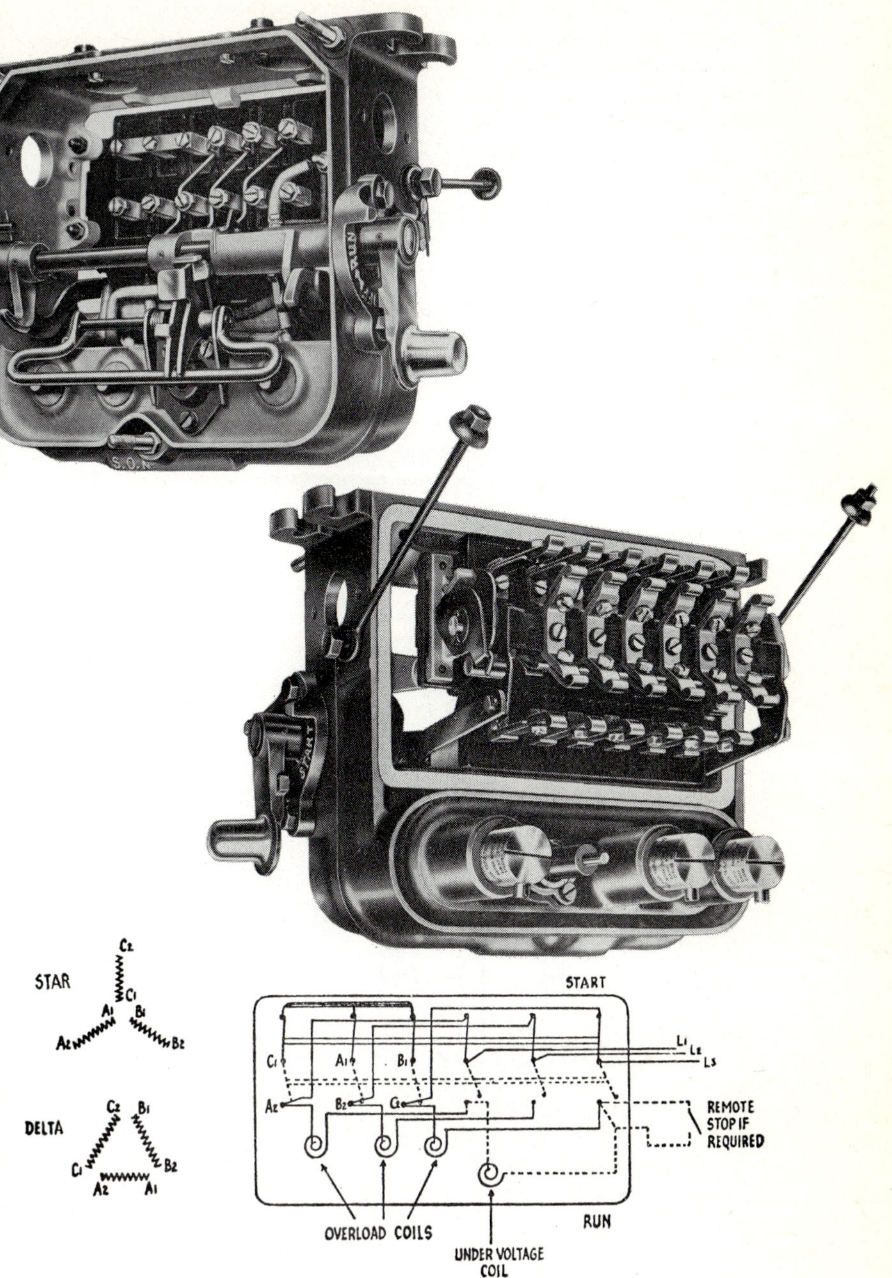

FIG. 135 *Hand-operated star-delta starter.*

a value that will reduce the current to a value only slightly in excess of 125 per cent of the full-load current of the motor, then it is very probable that the excess-current device will be operated.

No. There are six cables between the starter and the motor, two in each line, through which the current divides equally, therefore, cables with a lower current rating may be installed between the starter and the motor.

Because of bunching, however, and its effect on the temperature rise of the cables, they are not able to carry their rated current, and therefore cables must not be selected that can carry half the load current but fifty-eight per cent of the current.

122. *Describe, with the aid of a diagram, the operation of an auto-transformer starter and tabulate the ratios of the starting current and torque to the full-load current and torque of a motor which takes six times full-load current and twice full-load torque when started direct-on-line, if the transformer secondary winding has tappings at forty per cent, sixty per cent and eighty per cent of the line voltage.*

The auto-transformer starter (Fig. 136) comprises a change-over switch similar to that of the star-delta starter, and three auto-transformers. The auto-

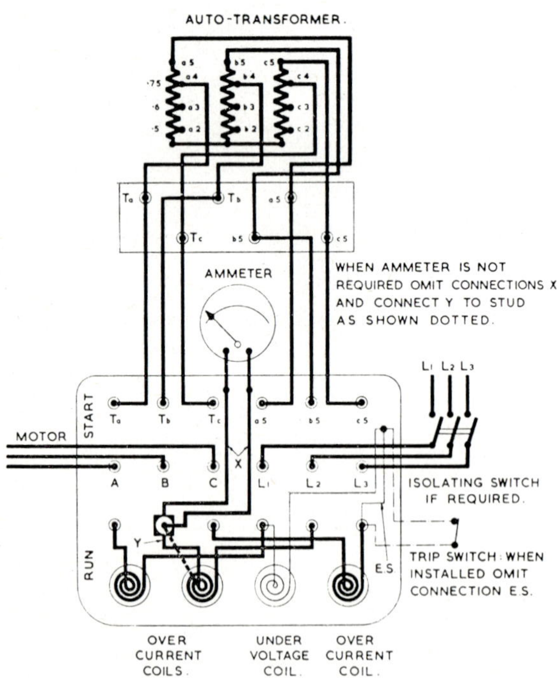

FIG. 136 *Hand-operated auto-transformer starter.*

transformers are tapped at three points, the tappings being connected respectively to the three terminals of three-way selector switch, a further connection of each switch being taken to the corresponding terminal on the starter. When the handle is turned to the start position, the six centre contacts are connected to the respective six top contacts. This connects the three supply lines to the respective three primary connections of the auto-transformers, while the motor is connected, via the selector switch, to the respective tapping on the secondary windings. Thus, a reduced voltage is impressed across the motor at starting.

When the motor has accelerated sufficiently, the handle is turned to the run position. As with the star-delta starter, if this is not done smartly, a catch prevents the handle from reaching the run position. The same catch prevents the motor from being started in the run position. The motor then receives its full voltage, the auto-transformers being entirely disconnected.

Although the voltage is reduced in proportion to the tapping on the auto-transformer, the starting current and torque are reduced in proportion to the square of the voltage across each winding. Because there are three tappings on the auto-transformers there is a wider range of starting currents and torques than those provided for by the star-delta starter, and the required torque output may be selected within prescribed limits to suit the required drive.

Table 4 illustrates the starting currents and torques for an auto-transformer starter with 40 per cent, 60 per cent and 80 per cent tappings for a cage-type three-phase motor with a starting current six times full-load current, and a starting torque of twice full-load torque when started direct-on-line.

Table 4

Tappings (%)	*Current* (*I*)		*Torque* (16lb. ft)	
100 (D.O.L.)	6		2	
80	3·64	× full-load current	1·28	× full-load torque
60	2·16		0·72	
40	0.96		0·32	

123. *In what circumstances is a shorting switch installed across the slip-rings of a three-phase, wound-rotor motor and what precautions must be taken where such a device is installed? What further interlocks are usually necessary? Give a wiring diagram of a starter suitable for starting such a motor, and from it construct a schematic diagram to illustrate the starting of the motor.*

The starting of a wound-rotor motor relies on the introduction of resistance in the rotor circuit during the starting period. This necessitates leads between the slip-rings of the motor and the resistance in the starter. Where the rotor current is large and the resistance is situated some distance away from the motor, some losses occur in the cables while the motor is in operation. These losses can be eliminated by installing a motor with a short-circuiting device across the slip rings. This device, by connecting a low resistance circuit in parallel with the rotor leads, bypasses the latter, so eliminating the power losses in the cables. Also, as the leads are only used during the starting period, they may be selected so that their size corresponds to half the full-load rating of the rotor current. With large motors, the saving in copper more than compensates for the increased cost of the motor.

Where such a device is installed, it is necessary to prevent the motor from being started with the slip-rings short-circuited. This is accomplished by connecting an auxiliary switch on the motor which is operated by a mechanism on the handle at the same time as the shorting device is closed. Thus, the contacts in the auxiliary switch are closed when the shorting device is in its normally open position. As the contacts are connected in series with the contact coil of the circuit-breaker, the breaker may be closed with the shorting device in the open position. Should the shorting device be operated before the motor is running normally, the contact coil remains unenergised and the circuit-breaker inoperative. When the motor is running, however, and the back e.m.f. of the motor is

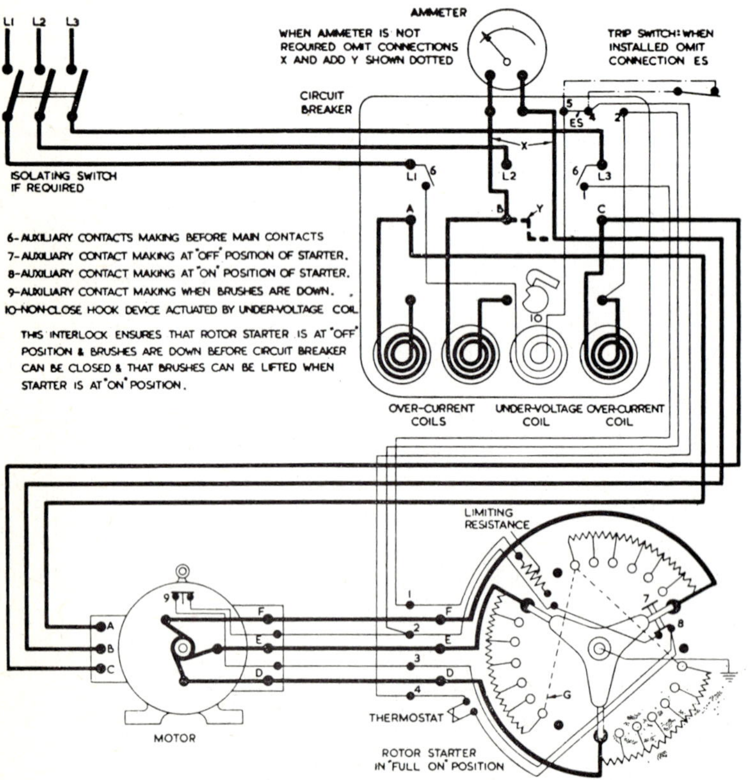

FIG. 137 *Wiring diagram for a three-phase wound-rotor starter.*

sufficiently established to keep the current in the motor to a safe value, the shorting device may be operated. As the auxiliary contacts on the motor are by-passed by a further pair of contacts in the circuit-breaker, the motor continues to run.

A further pair of contacts is also fitted in the starter resistance, the contacts being closed only if the handle is in position when all the resistance is in circuit. If any attempt is made to start the motor without full resistance in the rotor circuit, the contactor coil remains unenergised, and the motor fails to start. Where the resistors, which should be constructed as non-inductively as possible, are not designed for speed control of the motor, a further interlock is provided to open the contactor coil circuit, should the handle be moved away from its final position, so allowing resistance to be inserted in the rotor circuit. Fig. 137 illustrates the wiring diagram for such a motor and starter, which shows a further interlock being provided to disconnect the motor should the resistor become overheated. Fig. 138 illustrates the schematic diagram. Only by constructing such a diagram is it possible to really understand the operation of the motor and also how it facilitates any maintenance work which has to be carried out on the motor or starter.

When the handle of the circuit-breaker is closed. Switch 6 makes first, before the main breaker contacts. The circuit to the undervoltage coil is made, provided that the resistance arms are in their correct position, and, therefore, Switch 7

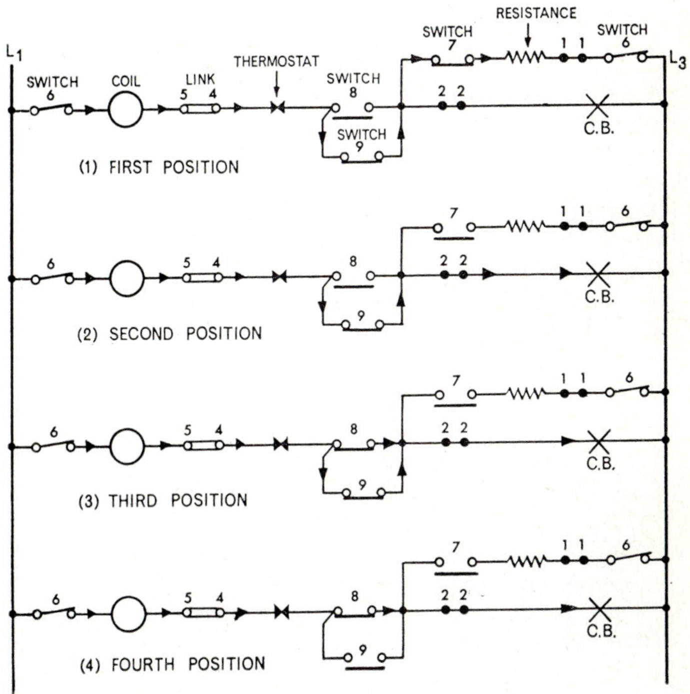

FIG. 138 *Schematic diagram for a three-phase wound-rotor starter.*

is closed. Switch 9 on the motor must also be closed, ensuring the slip-rings are not being short-circuited in the starting position. The undervoltage coil is thus energised, allowing the breaker to close, and the non-close hook device holds the breaker in the closed position as long as the undervoltage coil remains energised. This position is illustrated in Fig. 138 (1). In (2) the opening of Switch 7 does not de-energise the undervoltage coil as the latter is now connected in series with the main circuit-breaker to line L_3. Consequently, the resistance arm can be rotated, care being taken at each stage to allow the motor to accelerate sufficiently before the arm is moved to the next contact. (3) illustrates the situation where all the resistance has been cut out and switch 8 has been closed, thus closing a circuit in parallel with the motor switch (Switch 9). The final stage (4) shows how all the resistance has been cut out, the motor is running at its correct speed, the necessary interlocks have been closed and the short-circuiting device on the motor has been operated, so opening Switch 9. Notice that in no instance can the contactor coil become energised unless the short-circuit device is in its correct position, that is, Switch 9 is closed until the final stage is reached. It should also be noticed that if the resistances become over-heated at any stage in the sequence of operations, the thermostat is operated, so de-energising the undervoltage coil.

124. *How may the speed control of three-phase, induction motors be varied?*

The speed of the cage-type motor may be found from the expression, $N = 60f/p$ where f is the frequency of the supply, p the number of pairs of poles, and N is the synchronous speed of the motor. As most three-phase motors operate from a standard frequency supply of 50 Hz, the squirrel-cage motor is virtually a constant speed machine, once the number of poles has been decided upon. The following table gives the synchronous speed of cage-type motors with varying numbers of pole pairs.

Table 5

No. of pole pairs	*Synchronous speed (rev/min)*
2	3000
4	1500
6	1000
8	750
10	600
12	500
16	375

Considering the above table it can be seen that with a standard frequency supply, it is impossible to obtain a motor with a speed greater than 3000 rev/min. Also, no intermediate speeds can be obtained between the values given in the table. With some types of drives, however, it is an advantage to have the motor operating at two or even four different speeds. This is accomplished by so arranging the windings that they may be connected so as to give two values of the number of pole pairs. Where four speeds are required the motor is usually wound with two windings, each giving two values of pole pairs. Fig. 139 illustrates how a section of an eight-pole motor may be converted to a four-pole winding, the coils being connected in series for the lower speed and in parallel for the higher speed.

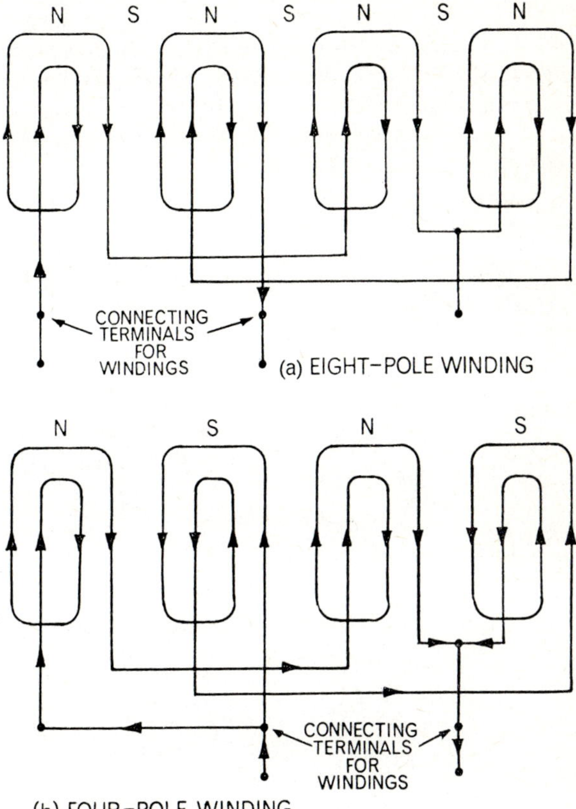

FIG. 139 *Illustrating speed control by pole changing.*

In many large industrial installions, motors, especially low-rated machines such as portable electric drills, are required to operate at speeds considerably higher than those obtainable on a 50 Hz supply. This necessitates the installation of a special machine termed a frequency convertor, driven by a motor served from the standard frequency supply. This enables motors to be driven at speeds in the order of 10 000 rev/min.

Another method which is often used for large motors is to install a modified form of induction motor known as the Schrage motor. The motor contains three windings, a primary winding, energised from the three-phase supply which is wound on the rotor, and the secondary winding which is wound on the stator and receives its energy inductively from the rotor winding. So far the description follows that of the wound-rotor motor with the stator and rotor windings in reverse. The Schrage motor, however, contains a third winding, termed the regulating winding, which is wound in the same slots as the primary winding but which is connected to commutator segments mounted on the same shaft. By making contact, through movable brushes, to the commutator segments, this winding is connected to the secondary winding, as shown in Fig. 140. There are two brush rockers, each holding three separate rocker arms, which carry the brushes. One end of each winding on the secondary winding is connected

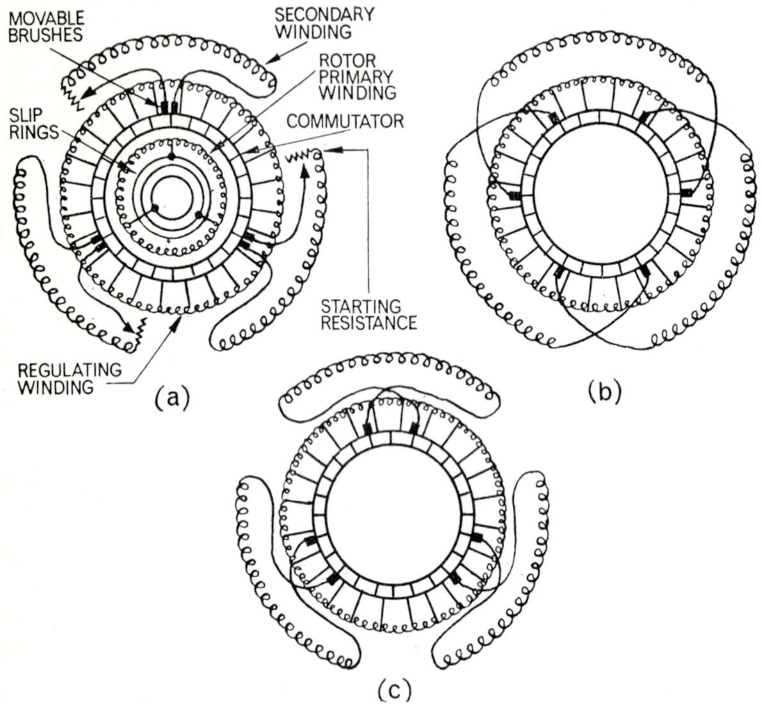

FIG. 140 *Illustrating speed control by a Schrage motor.*

to the brushes on one rocker arm, the other three ends being connected to the respective brushes on the other rocker arm. The rocker arms are geared so that they may be moved in opposite directions. When the brushes are in the position shown in (a), the regulating winding is short-circuited and the motor operates as the normal induction motor, running at slightly less than synchronous speed. When the brushes are in the position shown in (b), the induced e.m.f. is in the opposite direction to the e.m.f. induced when the brushes are in the position as in (c). Thus, by altering the relative position of the two brushes connected to each phase, any voltage between zero and maximum in either direction can be introduced into the secondary winding.

To understand how the brush position affects the speed of a motor, consider a wound-rotor motor with its slip-rings open-circuited and a voltmeter connected across them. At standstill, the reading is entirely dependent upon the transformation ratio between the two windings. If the motor is driven by some external means it will be found that the voltage varies with the speed. At standstill, it is at its maximum, at half the synchronous speed it is half the maximum voltage, and at synchronous speed, zero. At one and a half times synchronous speed it is again half the maximum voltage but of negative polarity (or of opposite phase rotation), and at twice synchronous speed it is again of maximum value but also of negative polarity.

Applying this reasoning to the Schrage motor, if at standstill, the maximum voltage on the regulation winding and stator winding are equal (say, at 120 V), with the brushes in one direction, the resultant voltage in the stator winding is zero ($120 - 120 = 0$). Hence no current flows in the windings and the rotor stays at rest. If the brushes are moved in the opposite direction to give maximum voltage (120 V) then the stator e.m.f. must be -120 V to produce a voltage balance ($-120 + 120 = 0$). Hence the motor must run at twice synchronous speed. Similarly, if the position of the brushes is arrange to give -60 V in one direction and $+60$ V in the other direction, then for voltage balance, the stator voltages would have to be $+60$ V and -60 V respectively, necessitating the motor running at half synchronous speed for the first brush position and one and a half times for the other brush position.

It is clear then, that different speeds may be obtained by varying the brush position so as to give voltage balance in the secondary circuit.

Although the speed is variable, once the brush position has been set, it runs at almost a constant speed. Normally, the motor is arranged to have ratios of 2:1 or 3:1 in either direction.

In comparison with motors whose speed regulation is obtained through resistance in the rotor, the efficiency is very high, especially at speeds above the synchronous speed. The power factor is also very high, approaching unity for high values of torque. At the synchronous speed setting, the power factor is similar to that of the normal induction motor.

Starting is effected by moving the brushes to the ‘in line’ position, and using direct-on-line starting, or by inserting resistances in the starting circuit, as shown in (a).

Although the initial cost of the Schrage motor is far higher than that of the slip-ring induction motor, the extra cost is soon recovered where frequent speed changes have to be used, because rheostatic losses are eliminated. Its chief applications are therefore for drives where frequent speed changes are required between those catered for by the two- or four-pole changing motor, such as for calendars, printing presses, pumps, spinning frames, paper-making machines, high-speed passenger lifts, etc.,

125. *Explain the term 'single-phasing' of three-phase star- and delta-connected motors. Describe two methods of preventing single-phasing.*

Single-phasing of three-phase motors occurs when one of the line connections between the motor and the supply is disconnected leaving the motor to operate on only two phases (Fig. 141). When the motor is delta connected, the three windings are connected across the two phases, but two of them are connected in series while the third winding is connected in parallel with the other two. Thus the current divides through the two circuits in inverse proportion to their impedances, so, although the currents in the two windings in series will increase, the current in the single winding will proportionally increase even more. If the mechanical load is sufficiently high, there is a tendency for the heavily-loaded winding to burn out if the single-phasing is maintained, especially if the overload is set on the high side.

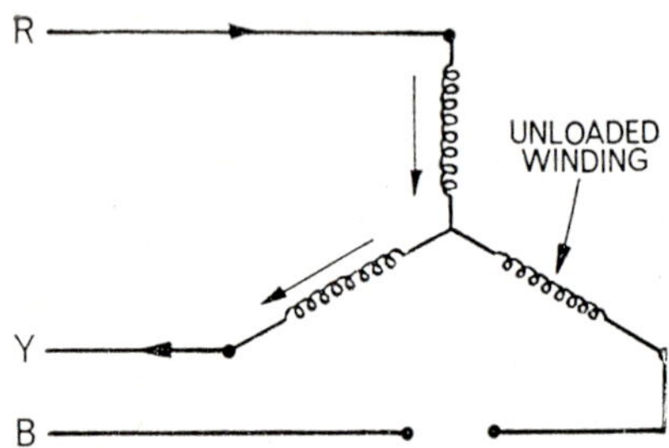

FIG. 141 (a) *Illustrating the effect of single-phasing.*

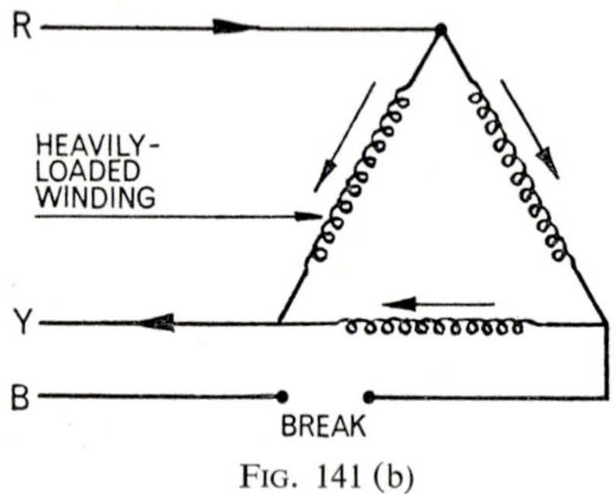

FIG. 141 (b)

When the motor is star-connected, two windings are connected in series across two phases. In this case, the third winding is disconnected and therefore is unloaded. Thus, when single-phasing occurs in a star-connected motor, as the current in the series circuit flows through the overload coils, normally these should be sufficient to protect the motor from burning out. But if they are set on the high side, or the incorrect grade of oil has been used, there is a possibility of both windings in the series circuit burning out.

Should the single-phasing occur before the motor is started, when the motor is connected to the supply, it will start to hum or try to run slowly. Should the single-phasing occur while the motor is running, then the motor will continue to run on the other two phases. If it continues to run indefinitely, a 'burn-out' is very probable.

The most common cause of single-phasing is a ruptured fuse, especially when direct-on-line starting is employed. Fuses of the rewirable type are very prone to deterioration because of the high currents taken by the motor when being started. Another cause is faulty contact in the associated switchgear, the inductive circuit causing arcing at the contacts when the circuit is being broken.

One method of preventing continued single-phasing is by incorporating a thermal overload device in the carcase of the motor which disconnects the motor from the supply when the temperature reaches a dangerous level.

FIG. 142 *Single-phase preventor.*

A second type of single-phase preventor (Fig. 142) contains three thermal strips which are heated by the respective currents flowing in the lines, and an auxiliary switch, the terminals of which are connected in series with the contactor coil. The single-phase preventor differs from the normal excess-current device in that the thermal strips are held between two hinged insulated bars, the bars themselves being held together by a spring, and the operating mechanism is actuated when the two bars are separated beyond a pre-determined limit. When current is flowing through the thermal strips, they bend upwards, causing the upper bar to be lifted. The lower bar, being held to the upper bar by the spring, has to follow this movement. When both bars have moved through a prescribed distance, the lower bar strikes against a stop, so retarding its movement. If excess current is maintained in the circuit, the upper bar continues to move upwards, until the two bars are separated sufficiently to actuate the tripping mechanism.

Although the stop is not designed to carry current, it is constructed of bimetallic strip with the same characteristics as the thermal strips, so that the relay is independent of ambient temperature, and the tripping current value is the same irrespective of temperature variations.

When single-phasing occurs, the thermal strip associated with the open-circuited line cools and bends downwards. Since current is still flowing through the other strips, they continue to move upward, and the bars separate until the tripping mechanism is actuated. The motor is therefore suitably protected against overloading and single-phasing.

126. *State the types of fault which occur in three-phase induction motors and say what steps should be taken to locate them.*

(1) *Locked rotor.* Should the rotor be unable to move, the belt or coupling should be removed and the bearings tested for wear by measuring the air gap with feelers, to ensure that the rotor is not rubbing on the stator laminations.

(2) *Motor overloaded.* If the motor starts satisfactorily with the load disconnected but not when connected, the starting torque is insufficient. If the motor and machine operate satisfactorily under running conditions, yet the motor still fails to start, the starting torque is again insufficient, and the load must be reduced or a larger motor installed. If auto-transformer starting is employed, the required increase in torque may be obtained by using a higher tapping on the transformer. It is possible, however, that the belt tension may be too tight, and adjustment of the belt should be tried before other methods of improvement are adopted.

(3) *Low voltage at motor terminals.* Low voltage at the motor terminals, caused either by a fall in the supply voltage or by excessive voltage drop in the cables serving the motor, reduces the capacity of the motor in proportion to the square of the voltage reduction across the motor. There may be, therefore, insufficient torque to start the motor. A voltmeter used progressively backwards will quickly detect if the drop in voltage is due to supply failure or to voltage drop in the cables.

(4) *Wrongly connected windings.* Wrongly connected windings may be found in either star- or delta-connected motors, where one of the windings may be connected in reverse order to the other two. Fig. 143 (a) illustrates vectorially the effect when the C winding is reversed in a star-connected motor, and Fig. 143 (b) the effect in a delta-connected motor. To remedy this, every connection must be checked between the starter and motor.

If the markings have been obliterated, or possibly the connections of the three windings themselves have been interchanged, they should be checked as follows. First, check through with a continuity tester or Avometer. Mark three ends A_1, B_1 and C_1 and the other respective three ends A_2, B_2 and C_2. Connect A_1, B_1 and C_1 in a star formation and the other three ends to the supply. The

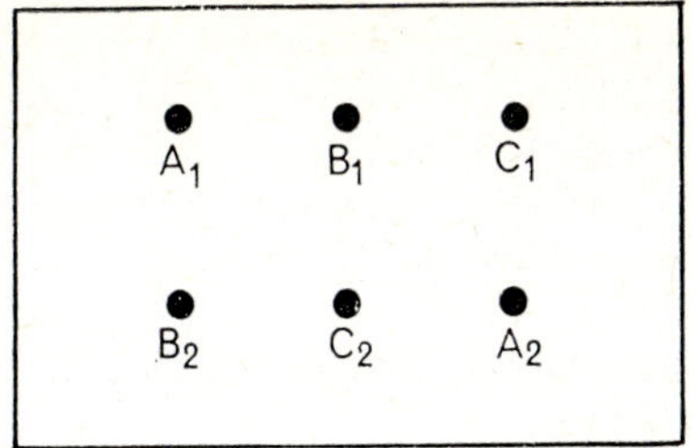

FIG. 143 (a) *Illustrating the effect of reversed winding and correct terminal markings.*

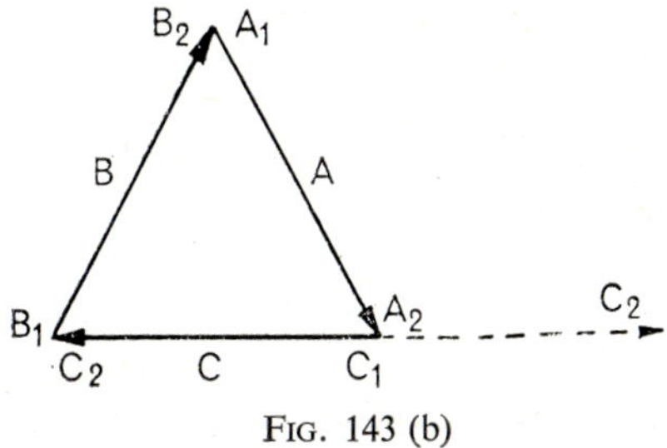

FIG. 143 (b)

motor should run if it is correctly connected, but if there is one reversed winding, the motor will try to start and will make a humming noise. Disconnect the motor and reverse one winding (say A_1 and A_2). If the motor still hums and refuses to start, restore A_1 and A_2 to their original positions, and repeat with the other two windings until the motor starts satisfactorily.

When the three windings have been correctly indentified, the correct connections should be made to their respective terminals on the terminal block, and if the markings have been obliterated, the six ends of the windings and the terminal connections should be marked as shown in Fig. 143 (c).

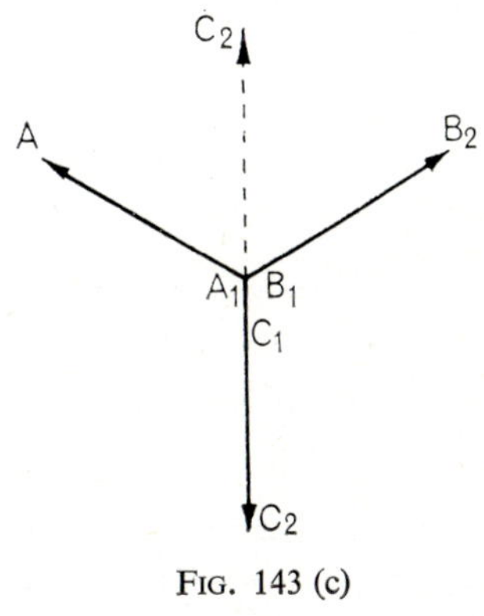

FIG. 143 (c)

(5) *Defective windings.* These may be found either by placing an ammeter in each winding, in which case an appreciable difference would be noticed in the reading if one winding is faulty, or alternatively by measuring the resistance of each winding with a Bridge Megger.

A defective rotor may produce: an increase in slip; noise; vibration; uneven starting torque; and variation of stator current on load. The squirrel rotor is virtually indestructible, but when all other possible causes of trouble have been eliminated, the motor should be stripped and the rotor examined for faulty welding or bar connections.

Rotor defects in a wound-rotor motor may be checked by means of a voltmeter. With the rotor resistance disconnected and the stator connected to the supply, the rotor is locked and a voltmeter is connected across each pair of slip-rings in turn. Serious imbalance of the readings will indicate a faulty winding. Alternatively, the resistance of each winding may be measured by means of a Bridge Megger, and again they should balance if the windings are in order.

(6) *Open-circuit in rotor external circuit.* This may be checked by measuring the resistance of each resistor which should all be approximately of the same value. A quick check is to bridge the slip-rings while the motor is running. With the starter in the 'full-on' position and any part of the rotor resistance defective, there should be a marked improvement in the running of the motor. Any flashing across the brush-holders would definitely indicate a fault in the external circuit.

(7) *Open circuit.* For single-phasing, see question **125.**

127. *Why is the single-phase induction motor not normally self-starting? What device must be installed in order to make the motor self-starting and what precautions must be taken when installing such a device?*

If we consider the two-pole winding shown in Fig. 144, when current is flowing through the winding, the alternating flux established by the winding induces currents in the rotor, which in turn produce their own magnetic field, the poles of which are diametrically in line with the poles established by the stator field. Hence, no torque is produced.

If the rotor is turned, the rotor conductors cut the stator magnetic field which induces e.m.f.s in them so causing currents to flow round the closed circuit. These currents produce a magnetic field, the poles of which are mid-way between the stator poles. Also, since the rotor is highly-inductive, the rotor currents lag behind the induced e.m.f.s by almost 90°. Thus, the rotor magnetic field pulsates almost 90° out of phase with the stator field, and because these two fields are at right angles, a rotating magnetic field is produced which maintains the direction of rotation.

It is necessary, therefore, to install an auxiliary winding, referred to as the starting winding, in order to simulate the running conditions while the motor is

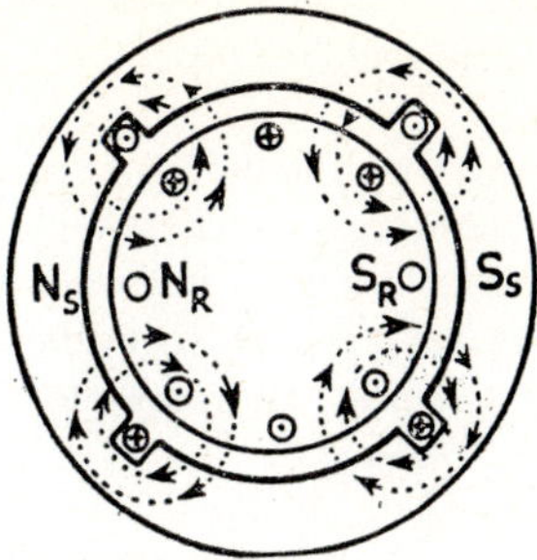

FIG. 144 *Illustrating why no starting torque is produced without winding starting.*

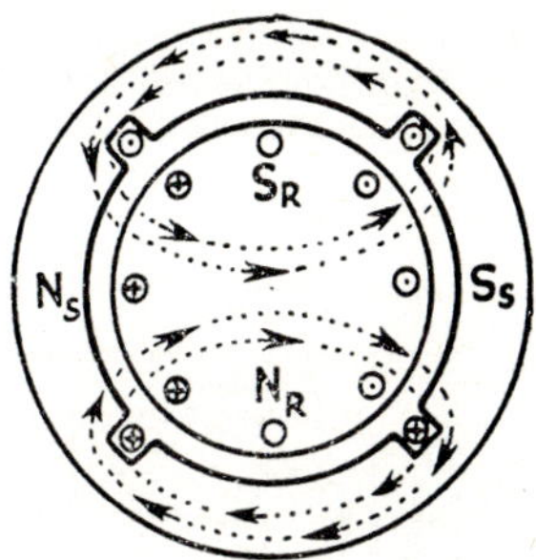

FIG. 144(b)

being started. This is placed mid-way between the main windings but as the torque is also proportional to the sine of the angle of phase difference of the currents in the two windings, it is necessary to have some means of producing this phase difference between the two currents.

In the normal split-phase motor, this is accomplished by making the starting winding highly resistive but of low reactance, and the main winding of similar reactance but low resistance. The phase difference, however, is not very large, and therefore a relatively poor starting torque is produced with this type of motor. If, however, a capacitor is connected in series with the starting winding, a greater phase difference between the two currents is produced, giving a much higher starting torque.

As the auxiliary winding is only used for starting the motor, it is only lightly wound and would burn out if left in circuit continuously. For small motors it is usual to incorporate a centrifugal switch which disconnects the starting winding when the motor reaches approximately 70 per cent of its rated speed. Larger motors, and even some small motors, may be arranged to have the winding disconnected externally.

128. *Describe the methods which may be used for starting single-phase split-phase motors and state the disadvantage of one of the methods.*

The most popular method, which is used for the majority of fractional kW motors, is to incorporate a centrifugal switch mounted on the shaft of the motor. The switch has normally-closed contacts, but when the motor has accelerated to about 70 per cent of its normal speed, the flying mechanism opens the contacts, disconnecting the starting winding (and capacitor if fitted). The method is illustrated in Fig. 145.

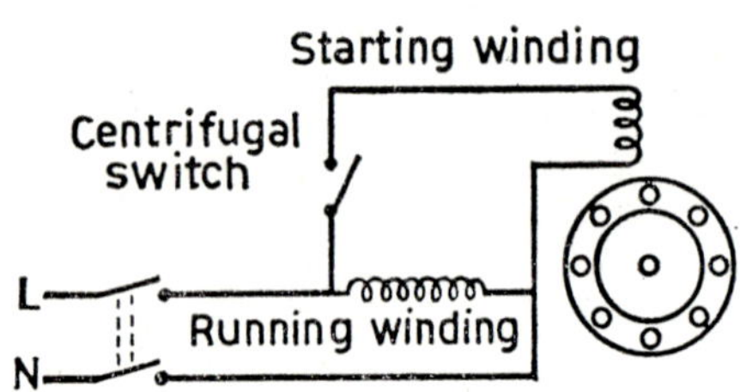

FIG. 145 *Starting a single-phase motor with a centrifugal switch.*

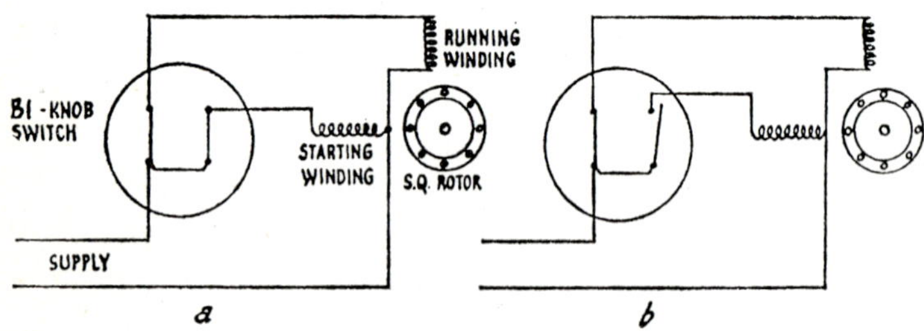

FIG. 146 *Starting a single-phase motor using twin-knob switch.*

Another method is by using a bi-knob switch. Fig. 146 illustrates the method, and it can be seen that the switch contains two sets of contacts, each controlled by its own dolly, and so arranged that both dollies are closed together. When the finger is released, a spring causes one of the switches to open, disconnecting the starting winding from the supply.

It is now necessary for virtually all motors to have undervoltage protection and therefore both the above methods must also incorporate such protection. Fig. 147 illustrates a starter which combines undervoltage and excess-current protection, and a phase-splitting device all in the same container.

For larger-powered motors, it is often necessary to install a rheostatic starter. Thus, the starter contains a graded resistor which is cut out in sections as the handle is rotated. The resistor is connected in series with the running winding, and the starting winding is connected so as to receive full voltage during the

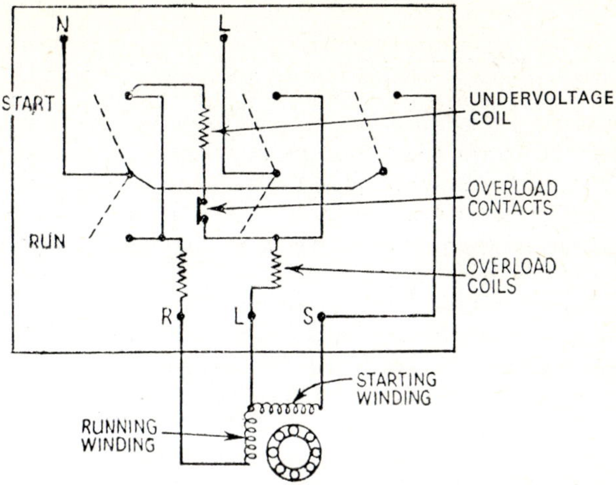

FIG. 147 *Single-phase starter with undervoltage and excess-current devices.*

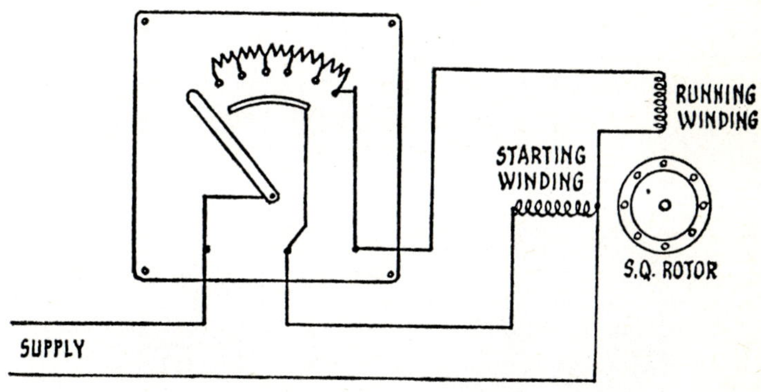

FIG. 148 *Rheostatic starting of a single-phase motor.*

starting period, but is disconnected just prior to the last section of the resistor being cut out. Although not included in Fig. 148, undervoltage and excess-current devices must also be incorporated in the starter.

Where a centrifugal switch is employed for disconnecting the starting winding, there is a possibility that, should the motor speed be substantially reduced, the centrifugal switch returns to the off-load position which closes the switch in the starting winding. The winding is thus re-connected to the supply, and if the fall in speed is maintained, the winding becomes overheated and eventually burns out.

129. *Describe the differences in the operation and performances of split-phase and capacitor-start single-phase motors, and state for what duties each is respectively suitable.*

Because of the method used for obtaining the phase difference in the split-phase motor, the currents, and therefore the magnetic fields, have only a low phase difference between them, and, since the torque is proportional to the sine of the angle of phase difference between the two currents the starting torque of such a motor is relatively low when switched directly on to the supply, being approximately 1·5 times the full-load value, while the current at start is approximately 7 times the full-load current. The motor then is only suitable for drives with a low starting inertia, and where the number of starts per hour is limited. The efficiency and power factor are much lower than those of a correspondingly rated three-phase motor, averaging 65 to 75 per cent and 0·7 to 0·8, respectively.

This type of motor is suitable for fans, blowers, centrifugal pumps, sewing machines, lathes, etc., where the starting inertia is low.

As this type of motor is used extensively in domestic installations, silent running is essential, and, therefore, in addition to the fitting of sleeve bearings instead of ball bearings, resilient mounting (Fig. 149) is often adopted which ensures that the motor is even more silent and free from vibration.

The addition of the capacitor in the starting winding circuit of the capacitor-start motor ensures that the phase difference between the currents in the two windings is considerably higher than that of the split-phase motor, hence the

FIG. 149 *Resilient mounting.*

starting torque is also considerably higher, being approximately 4·5 times full-load torque. Also the presence of capacitance in the circuit results in a reduction in the starting current in the order of approximately 4 times full-load current. Although the starting of the motor is improved by the addition of the capacitor, as it is disconnected with the starting winding, the running performance is similar to that of the split-phase motor.

This type of motor is suitable where the inertia of the load is high at start, as in drives for refrigerators, air-compressors, pumps, etc.,

130. *In what way does the capacitor-start-and-run motor differ from the capacitor start motor, and why is it necessary to incorporate two capacitance values within the former motor?*

In the capacitor-start motor, the capacitor is used only for starting, whereas in the capacitor-start-and-run motor, some of the capacitance is left in the starting winding circuit during the time the motor is running. By this means, not only has the motor a good starting torque, but the addition of the starting winding in the running position gives a greatly improved running performance, the motor operating somewhat similarly to the two-phase motor.

Because the low running current and high initial cost do not warrant its installation, the capacitor-start-and-run motor is seldom found in fractional kilowatt sizes, but it is very suitable for larger kilowatt motors because its higher power factor and efficiency result in an appreciable saving in the running costs, compared with the split-phase motor.

The necessity for the two values of capacitance is because at start, the frequency of the rotor currents is the same as the supply frequency, hence the reactance value of the winding is high ($X = 2\pi fL$). Therefore, a high value of capacitance is necessary to neutralise the inductive reactance. As the motor accelerates, the frequency of the rotor current falls to a low value, hence the inductive reactance also falls, necessitating part of the capacitance being disconnected to compensate for the reduction in the inductive reactance. Fig. 150 illustrates the wiring diagram for a capacitor-start-and-run motor.

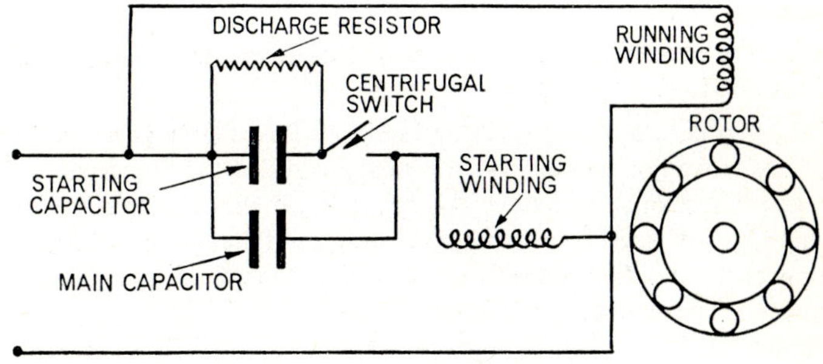

FIG. 150 *Capacitor start-and-run motor.*

H

14

Illumination

Illumination terms—illumination calculations—the lightmeter—hot-cathode fluorescent circuits—stroboscopic effect—cold-cathode lighting circuits—high-pressure mercury vapour lamp—planning illumination schemes.

131. *What is meant by the following illumination terms:* (1) *luminous intensity;* (2) *luminous flux;* (3) *illumination; and* (4) *brightness?*

(1) Light is a form of electromagnetic energy, therefore, if we wish to compare the amount of light energy emitted by one source with another, we compare what is termed the 'luminous intensity' of both sources. The unit of luminous intensity is the candela (cd after numerals) which is defined as the luminous intensity of such magnitude that the luminance of radiation of platinum maintained at a temperature of solidification (1773° C) is 0·60 cd/mm^2. The symbol is I.

(2) Luminous flux is the light radiation emitted in every direction from the light source. The unit is the lumen which is defined as the light radiation emitted per unit solid angle from a light source with uniform light intensity of 1 candela. (Note the solid angle or steradian is a unit of spherical measurement, there being 4π solid angles in a sphere.) As there are 4π solid angles in a sphere, the luminous flux emitted by a light source of 1 cd is $4\pi I$ lumens. The symbol for luminous flux is F.

(3) When a light source is emitting luminous flux, the flux eventually reaches the surface to be illuminated, but the effectiveness of the illumination depends upon the number of lumens the surface receives. It should also be clear that the number of lumens is also dependent upon the area of the surface to be illuminated. The unit of illumination is therefore defined as the unit flux per unit area (1 lux = 1 lm/m^2). The symbol for illumination is E.

Relating the above quantities to one another, if a light source of 1 cd is placed at the centre of a 1 m radius sphere, the light source is emitting 1 lm per unit solid angle or 4π lumens altogether. Each lumen is illuminating 1 m^2 of surface, the illumination value being 1 lux.

(4) Brightness may be defined as the ratio of the luminous intensity to the area of the light source. Thus, where the light source is concentrated in one small area, it can be harmful or even dangerous to the eyes if looked at directly, whereas if the light source is distributed over a large area, as in a fluorescent lamp, the light source becomes quite bearable to the human eye. All good lighting installations should be designed so that glare is reduced to a minimum. The symbol of brightness is B, the unit being the candela per unit area (cd/m^2).

132. *What is meant by the term 'cosine law' as applied to illumination problems?*

If a small screen is placed directly beneath a light source with its surface at right angles to the light source, the illumination received at the surface, in accordance with the inverse square law, is I/d^2. If the screen is rotated through 90° so that its surface is lying parallel to the light source, then, if we consider the light source as a point and neglect reflection, the surface of the screen would receive no illumination at all. This condition would exist no matter how near the screen was to the light source. In actual fact, the light source is never a true point and, as there are nearly always adjacent surfaces to reflect light, the surface would receive a certain amount of illumination.

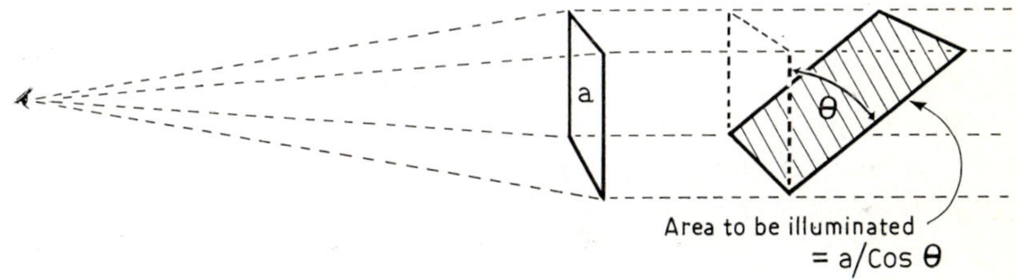

FIG. 151 *Illustrating the cosine law of illumination.*

If, in Fig. 151, the surface is turned so that it makes the angle θ with its original position, the area to be illuminated has been increased in the ratio of $1/\cos\theta$. The illumination is therefore reduced in the ratio of $\cos\theta$. Hence the illumination received at any surface relative to a light source may be found from the formula:

$$E = I\cos\theta/d^2$$

Note that this formula applies even when the surface is directly beneath the light source as, in this instance, θ is zero and therefore $\cos\theta$ is unity.

133. *What is meant by the term 'cos³ law' as applied to illumination problems, and show how it is derived?*

If, in Fig. 152, h is the height of the light source above the point A on the surface, d the distance of the light source to the point B, and θ the angle of incidence, then the illumination at B is:

$$E_B = \frac{I \cos \theta}{d^2}$$

but as $$\cos \theta = \frac{h}{d}$$

the formula can be written $$E_B = \frac{Ih}{d^3}$$

cross multiplying $$I = \frac{E_B d^3}{h}$$

Also, at point A $$E_A = \frac{I}{h^2}$$

cross multiplying $$I = E_A h^2$$

and therefore $$\frac{E_B d^3}{h} = E_A h^2$$

cross multiplying $$E_B = \frac{E_A h^3}{d^3}$$

as $h/d = \cos \theta$ $$E_B = E_A \cos^3 \theta$$

Thus, the illumination at any point on a level surface is proportional to the illumination at a point directly beneath the light source and the cube of the cosine of the angle of incidence.

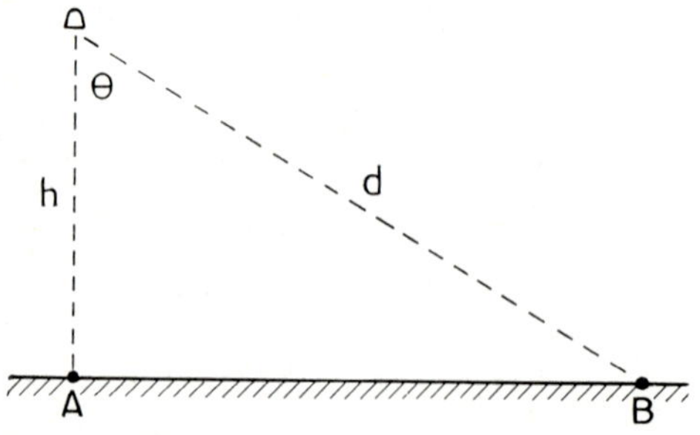

FIG. 152 *Illustrating the cosine³ law of illumination.*

134. *Describe the operation of a hot-cathode, switch-start fluorescent lamp circuit.*

In this type of lamp much of the light radiated is from the fluorescent powder contained on the inner surface of the tube, which is excited under the influence of the ultra-violet rays emitted by the discharge. This entails, therefore, the lamp operating at low pressure. The tube is comparatively large, and operates at much lower current density than the high-pressure mercury-vapour lamp. In addition to the mercury, the tube contains a small amount of argon gas to assist in initiating the discharge. Even with the addition of the argon gas, however, the lamp is not self-starting, and therefore special starting devices are necessary.

These consist of two electrodes contained in the tube, each of which has two parts electrically connected to each other. One part consists of a tungsten filament

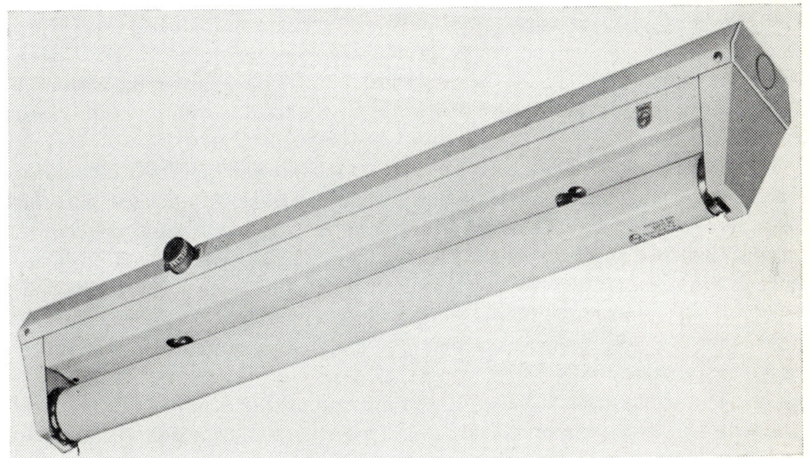

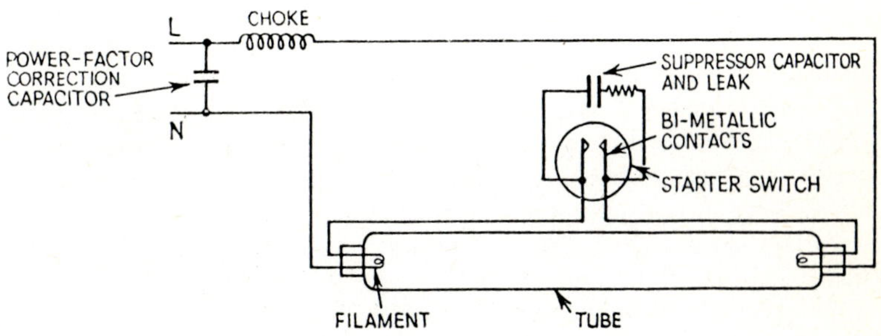

FIG. 153 *The hot-cathode glowlamp fluorescent lamp circuit.*

coated with oxides which readily shed electrons when heated. The other part consists of a metal strip or fin which serves as the anode, when the other filament is operating as the cathode. Even then, this is not sufficient to initiate the discharge, and, therefore, a choking coil and a starter switch are incorporated in the circuit. The function of the starter switch is to disconnect the choke and so impress a high-voltage surge across the tube.

The glowlamp type of starter switch (by far the most commonly used) comprises a glass envelope containing helium gas and two bimetallic contacts which are normally open (the complete circuit being shown in Fig. 153).

When the circuit is connected to the supply, current flows through the chokes through the filaments of the lamp and through the gas in the starter lamp. At this stage, most of the potential drop is across the lamp, due to the low current in the circuit. Ionisation starts to take place in the immediate neighbourhood of the electrodes, and simultaneously the glowlamp starts to warm up. As the temperature of the glowlamp increases, the contacts move towards each other eventually touching, so causing the glow in the starter lamp to be extinguished. The glowlamp starts to cool and eventually the contacts re-open, opening the choke and lamp circuit. This results in a voltage of approximately 900 V being impressed momentarily across the tube which, in conjunction with the ionisation produced by the filament heating, is normally sufficient to start the discharge through the tube.

The choke is now required to perform its second function, that of limiting the current in the circuit. Thus, as the current increases, so does the potential drop across the choke. Therefore, the potential across the lamp is reduced, until a point of balance is reached where the potential across the lamp is just sufficient to maintain the discharge at its correct value (approximately half the supply voltage). When the tube is functioning correctly, the discharge current is sufficient to keep the electrodes at the right temperature, so rendering the starter circuit inoperative.

Owing to choke control having to be incorporated, the power factor of the circuit is low, being approximately 0·5. Therefore a capacitor is necessary to improve the power factor to about 0·9.

The great advantage of the fluorescent lamp is its high efficiency value when compared to that of the tungsten-filament lamp, the efficiency being in the order of 50 lm/W, more than three times the efficiency of the average tungsten-filament lamp. This is slightly offset by a power loss in the choke.

Another advantage is that different colour effects, even simulated natural daylight, can be obtained by varying the mixtures of the fluorescent powders applied to the inner surface of the tube. Also, as the luminosity is spread over a relatively large surface area, glare is greatly minimised.

A further advantage is that, if installed correctly, fluorescent lighting is virtually shadowless.

Although the installation cost is relatively high compared to a tungsten-filament lighting scheme, the lower operational costs and the benefit of more efficient lighting, somewhat compensates for this.

135. *What is a disadvantage of a switch-start fluorescent lamp circuit which is overcome in the instant-start circuit?*

One of the disadvantages of the switch-start fluorescent lamp circuit is that a period of time must elapse before the discharge is operating normally. The disadvantage is overcome in the instant-start circuit of the hot-cathode type (Fig. 154) by directly heating the filaments from the secondary winding of an auto-transformer. In addition, an earthed-metal strip, placed lengthways along the tube, terminating in the vicinity of the electrodes, assists in the ionisation of the gas which, in combination with the direct heating of the electrodes, produces virtually instant starting. No high-voltage surge is necessary and the choke is used only as a current-limiting device.

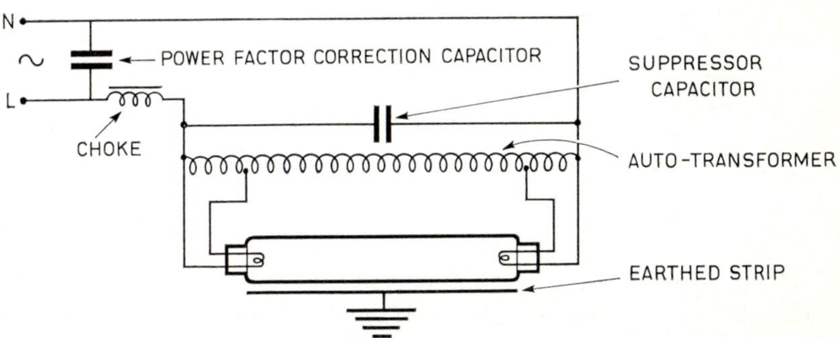

FIG. 154 *The instant start hot-cathode fluorescent lamp circuit.*

At start there is very little potential drop across the choke, most of the potential being across the transformer and the tube. As current flows through the choke, the potential across it rises, while the potential across the lamp falls until a balance is obtained, when approximately half the supply voltage is lost across the tube.

The power factor is again low because of the presence of the choke, so necessitating the incorporation of a capacitor. Also with both the switch-start and instant-start circuits, suppressors capacitors are necessary to suppress interference with radio waves, the capacitor being connected across the starter lamp in the switch-start circuit and across the auto-transformer in the instant-start circuit.

136. *Describe how a fluorescent lamp circuit may be modified to operate on d.c. supplies.*

On d.c. supplies, the choke becomes inoperative as a current-limiting device, and, as current limitation is necessary, a resistor must be incorporated in the circuit. The choke, however, is still normally necessary to assist in the starting of the lamp. Also, because, the mercury tends to migrate towards one end of the tube when the lamp has been operating for a time, a reversing switch is necessary to reverse the polarity so that the discharge conditions are equalised (Fig. 155).

Since a resistor is used as a current limiting device instead of a choke, relatively heavy power losses are incurred in it, the efficiency being much lower than that of the inductor operated lamp and not appreciably higher than that of the tungsten-filament lamp.

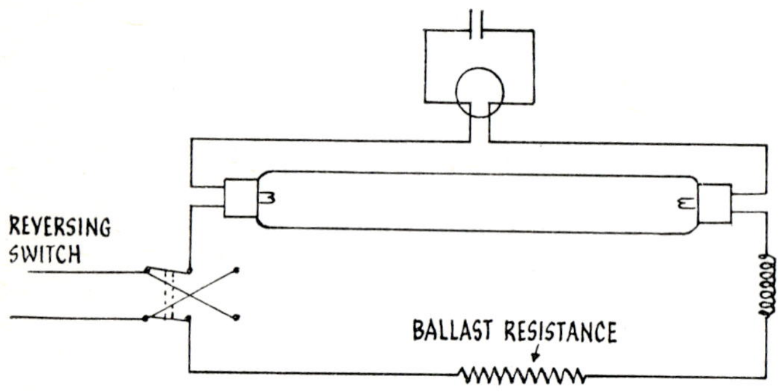

FIG. 155 *Glowlamp circuit modified for use on d.c. supplies.*

137. *Describe the stroboscopic effect and state how its effect may be minimised.*

The stroboscopic effect occurs when an object which is moving appears to be stationary. On filament lighting, although the current passes through the zero point twice every cycle, on standard frequency supplies, the light output (which is due to the heating of the filament to incandescence) has no time to diminish before the current has passed through the zero point and is approaching its maximum value again. Thus, the stroboscopic effect is not apparent on filament lighting installations.

With discharge lighting, the light is actually extinguished twice every cycle. Thus, considering a spoked wheel, if one of the spokes occupies exactly the same position as the preceding spoke at similar positions in the half cycle of the waveform (possibly when the current is passing through the zero point), then the wheel will appear to be stationary. The effect is not very pronounced with fluorescent lighting, as the powder tends to maintain its fluorescence even when the current is passing though its zero point. On high-pressure mercury-vapour lighting installation, the effect is very noticeable and can be dangerous to operatives.

One method of minimising the stroboscopic effect on three-phase supplies is to connect adjacent lamps to different phases, so that all the lamps in the immediate neighbourhood of any rotating machinery are not passing through the zero point in the cycle at the same time.

With fluorescent lighting, use is made of the twin-fitting circuit (Fig. 156).

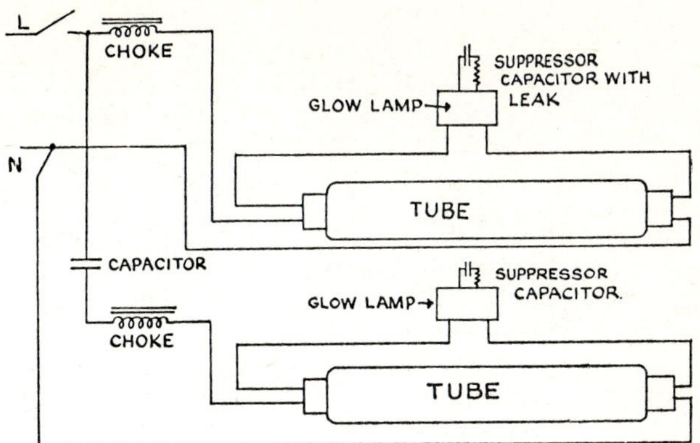

FIG. 156 *Twin fluorescent lamp circuit for overcoming the stroboscopic effect.*

Only the one main capacitor is used, and this is of sufficient capacity to improve the power factor of both lamp circuits to approximately 0·9. The capacitor is connected in series with one lamp circuit and, as this circuit therefore takes a leading power factor, while the uncorrected lamp circuit takes a lagging power factor, the currents in the two lamps are never passing through the zero points in the current cycle at the same time, so minimising the stroboscopic effect.

138. *Describe the operation of the cold-cathode fluorescent lamp circuit, and state its advantages and disadvantages.*

In the hot-cathode type of fluorescent lamp, because of the necessity of filament heating, a certain potential is lost across the cathode. This is referred to as the cathode fall, and for this type of lamp is approximately 5 V, so that the efficiency at the normal supply voltage is not greatly reduced by this loss.

In the cold-cathode fluorescent lamp, there is no filament heating and the electronic emission is literally snatched from the cathode by applying a relatively high potential near to the cathode in the form of an electrostatic field. This potential has to be in the order of 200 to 300 V, thereby rendering the lamp unsuitable for normal supply voltages. As this potential is constant and irrespective of the length of the tube, the efficiency of the lamp may be increased by making it longer. This, in turn, requires a higher potential across the circuit so that the cold-cathode lamp is essentially a high-voltage lamp, the efficiency then being comparable with, but slightly lower, than that of the hot-cathode lamp.

One common arrangement (Fig. 157) shows three lamps, each 3m long, which are connected in series with the secondary windings of two transformers, the common connection of the two secondary windings being earthed. The transformers are of the high-reactance type, and so act as current-limiting devices, the voltage falling from the starting voltage of approximately 3600 V to the running voltage of 1900 V. In the type illustrated, the tubes are rated at 70 W, the current in the tube being 120 mA and the current taken from the mains being 1·3 A. The reason for this relatively high current is because of the abnormally low power factor of the circuit, which is approximately 0·3. Thus a capacitor is required to improve the power factor of the circuit to about 0·85.

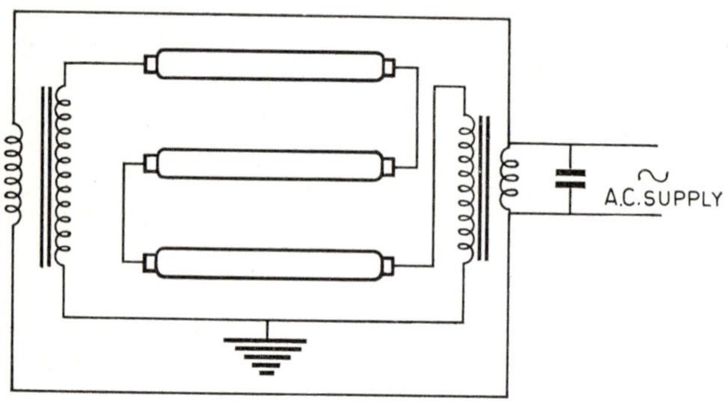

FIG. 157 *Cold-cathode fluorescent lamp circuit.*

One advantage of this type of lamp is that it is instant starting. Also because no specially treated filaments have to be incorporated, the life of the lamp is very long, in the order of 10 000 hours.

The disadvantages are the high voltage necessary to operate the lamp and also that it is not so compact as the smaller hot-cathode type and therefore not so adaptable.

139. *Describe the construction and operation of the high-pressure mercury vapour lamp and state why it is necessary to incorporate a choke.*

This type of lamp operates at high pressure, the pressure being between 1 and 2 atmospheres when the lamp is operating normally. It consists of an inner and outer chamber, the space between being partly evacuated. The inner chamber contains a small quantity of mercury and argon gas, the latter being necessary to facilitate the starting of the discharge. The inner chamber is constructed of hard quartz glass, and the lamp is operated vertically so that the discharge does not come into contact with the glass. The outer chamber is of ordinary glass construction. Its purpose is to absorb any ultraviolet rays which might penetrate the inner chamber while the lamp is in operation and also to provide an evacuated space to prevent undue loss of heat and so keep the temperature of the inner chamber reasonably constant. The inner chamber also contains the two main electrodes which consist of tungsten filaments surrounding electron-emitting cathodes. No heating of the filaments is necessary, and for starting purposes an auxiliary electrode is placed adjacent to one of the main electrodes. This consists of a fine wire and resistance (Fig. 158), the value of the latter being in the order of 50 000 Ω.

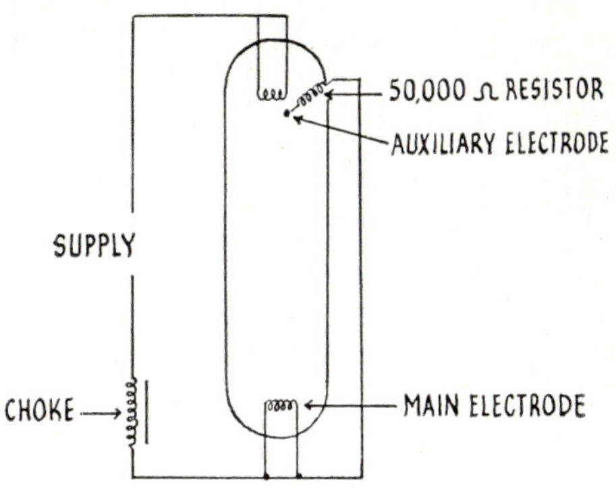

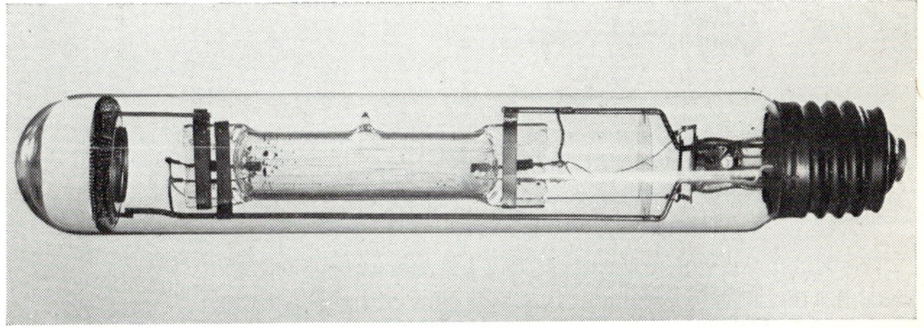

FIG. 158 *High-pressure mercury vapour lamp circuit.*

When the lamp is connected to the supply a small discharge is initiated between the auxiliary electrode and main electrode adjacent to it, caused by ionisation of the argon gas. This discharge is communicated to the argon gas between the two main electrodes, and the temperature of the lamp starts to increase. As the temperature increases, the mercury vaporises, so increasing the pressure inside the inner tube, the luminous column becoming brighter and narrower until, when the mercury is full vaporised, the lamp attains its full brilliancy. The voltage across the lamp rises, during this period, from approximately 20 V to 120 V. When fully operating, the temperature of the lamp is about 600°C.

The purpose of the choke is to prevent the current from becoming excessively high. This reduces the power factor of the circuit to approximately 0·5, therefore a capacitor is also incorporated to improve the power factor to about 0·9. One advantage of this type of lamp is its relatively high efficiency, which is in the order of 45 lm/W, representing an efficiency of almost three times that of the average filament lamp.

One disadvantage is that the lamp takes a relatively long time to attain its maximum brilliancy, and, should the lamp become disconnected from the supply, the pressure inside the lamp has to fall before the discharge is restarted, increasing the time taken for the lamp to be restored to its original brilliancy. It is necessary, however, to operate the lamp at a relatively high pressure, otherwise much of the discharge would radiate rays below the visible wavelength (that is, ultra-violet rays).

Another disadvantage lies in the colour radiated by the lamp. Many people find this objectionable to work under, especially in non-industrial installations. If, however, reflecting surfaces are coloured cream, so that the blue rays are mellowed by the rays reflected from the surfaces, this objection largely disappears.

Another method of overcoming this disadvantage is to incorporate a tungsten filament in the same outer chamber as the discharge tube. As well as providing colour correction, the filament may also be used as a current-limiting ballast, so that no choke or capacitor is necessary.

Alternatively, separate tungsten-filament lamps may be mounted adjacent to the mercury-vapour lamps so providing colour correction.

140. *In planning illumination schemes what is meant by the terms 'coefficient of utilisation' and 'maintenance factor', and what factors must be taken into account when determining their values?*

The 'coefficient of utilisation' may be defined as the ratio of the light flux reaching the working plane to the total emitted light flux.

The 'maintenance factor' is a factor which is applied to account for the reduction in the light output due to dust and dirt accumulating on the lamp, the light fitting and the surrounding fixtures.

Thus the required lumen output is proportional to the required illumination, and the area to be illuminated, but inversely proportional to the 'coefficient of utilisation' and the 'maintenance factor'.

$$\text{Required lumen output} = \frac{E \times A}{\text{coefficient of utilisation} \times \text{maintenance factor}}$$

The coefficient of utilisation depends upon:

(1) The type of fitting to be installed. (Direct lighting schemes where most of the light is reflected downwards have a far higher coefficient of utilisation than indirect lighting schemes, where most of the light is directed upwards.)

(2) The room index which is determined from the dimensions of the room and the mounting height of the fitting.

(3) The reflection factor of the walls and ceilings.

Although the maintenance factor has a fairly constant value of approximately 0·7, the coefficient of utilisation, because of the wide variation in the above factors, has a very wide range from below 0·15 for indirect lighting schemes, to unity for certain direct lighting schemes. For this reason, tables are compiled giving the various utilisation factors for different types of fittings, and the room indices and reflection factors of the walls and ceilings.

141. *A small workshop* 20 *m by* 20 *m is to be illuminated at bench level to a value of* 150 *lux. The following lighting schemes have been proposed:*

(*a*) 150 *W tungsten-filament lamps giving* 13 *lm/W.*

(*b*) 80 *W fluorescent lamps giving* 40 *lm/W.*

Assuming the coefficient of utilisation of the room to be 0·6 *and the maintenance factor in each case to be* 0·8, *calculate the number of lamps for both methods.*

$$\text{Area} = 20 \times 20$$
$$= 400 \text{ m}^2$$

$$\text{Total lumen output required} = \frac{\text{illumination} \times \text{area}}{\text{coeff of utilisation} \times \text{maintenance factor}}$$
$$= \frac{150 \times 400}{0{\cdot}8 \times 0{\cdot}6}$$
$$= 125\,000 \text{ lm}$$

$$(a) \text{ Lumen output per lamp} = 150 \times 13$$
$$= 1950 \text{ lm}$$

$$\text{therefore number of lamps required} = \frac{125\,000}{1950}$$
$$= 64 \text{ lamps}$$

$$(b) \text{ Lumen output per lamp} = 80 \times 40$$
$$= 3200 \text{ lm}$$

$$\text{therefore number of lamps required} = \frac{125\,000}{3200}$$
$$= 40 \text{ lamps}$$

15

Heating

Methods of heat transfer—space heaters—space heating thermostats—space heating calculations—water heaters—water heating thermostats—heat-controlled iron—simmerstat control—faults on heating appliances.

142. *Describe the types of heater available for space heating, and state for what types of situation each would be best suited.*

Radiant heaters. Radiant heaters are normally of two types; (1) the firebar type and (2) the reflector type (Fig. 159). In the firebar type, the resistance, which may be of nickel-chrome, is laid in grooves consisting of a series of

FIG. 159 *Reflector type radiant heater.* (*Courtesy Belling Ltd.*)

parabolic wells in a block of fireclay. The resistance wire has normally a loading of 1 kW, and where higher ratings are required, 2 or 3 elements may be combined in the one fire. The heat output of this type of heater is approximately 60 per cent radiant heat, and 40 per cent convective heat.

In the parabolic reflector type of fire, the element consist of a rod of fireclay containing a spiral groove in which is wound, almost to the full length of the rod, the heating element. Alternatively, the element may be contained in a silica tube. The rod is then placed at the focal point of a highly polished reflector, so that the element, when hot, reflects heat in the form of a beam which is approximately 70 per cent radiated heat and 30 per cent convective heat. The element may have loadings of 500 W, 750 W or 1000 W, and where fires of higher rating are required, elements may be connected in parallel.

Radiant heaters are generally unsuitable for industrial installations. They are, however, satisfactory for domestic installation where they are used for the local heating of small rooms. They are also suitable for some non-domestic installations, such as church halls, where rapid heating is required for only short periods of time.

Convector heater (Fig. 160 (a) (b)). These normally consist of sheet-metal cases, which may be plastic finished for appearance, containing inlet and outlet openings at the bottom and top of the heater, both of which may contain louvres. The heating element is situated in the lower half of the heater, and consists of wire-wound elements operating at a dull or black heat. Cool air is drawn in through the bottom inlet, passed over the element then discharged through the top outlet as warm air. This then circulates around the room.

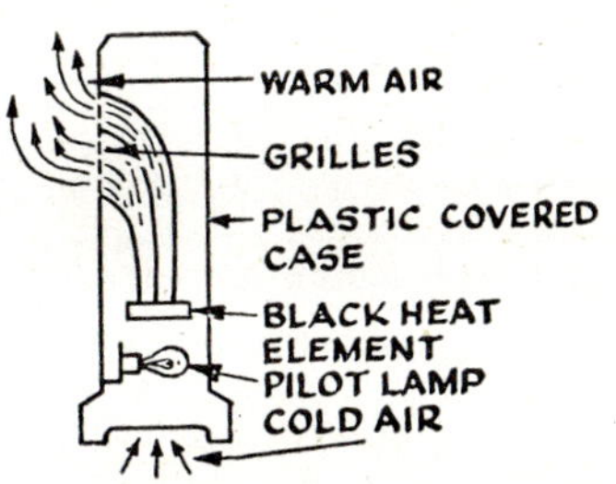

FIG. 160 (a) *Convector heater.*

The heater may be of a portable design or suitable for wall mounting, and for domestic use is normally obtainable in 1 kW, 2kW or 3 kW loadings. It is also suitable for thermostatic control.

Convector heaters are generally suitable for small installations where good overall heating is required. When used in conjunction with a radiant fire, they produce a very satisfactory means of heating small rooms.

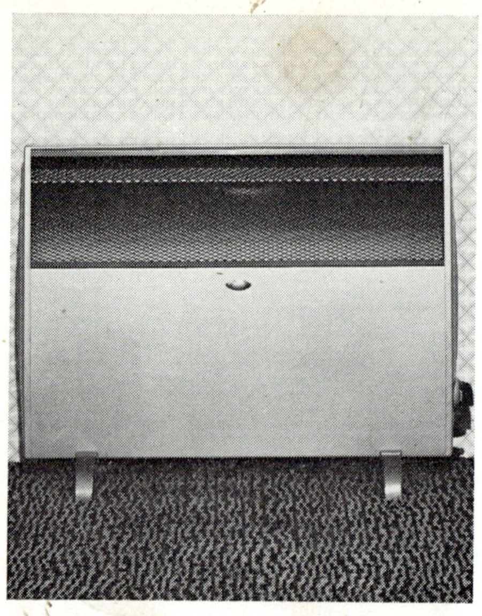

FIG. 160 (b) *Convector heater* (*Courtesy Belling Ltd.*)

Low-temperature tubular heaters (Fig. 161). Because of the complete absence of exposed heating elements, and also because of the relatively low surface temperature, tubular heaters are often used for local heating such as halls, stairways, etc. A typical heater may comprise a heavy-gauge seamless, round

FIG. 161 *Low-temperature tubular heater.*

steel tubing, approximately 50 mm in diameter, or it may be elliptically shaped, containing a coil of nickel-chrome wire interlaced on a former constructed of mica, fireclay or porcelain. The element operates at a dull heat, giving a surface temperature of approximately 95°C and is so interlaced that the connections both finish at the same end. The loadings may be 197 W, 263 W or 328 W per metre and lengths are generally up to 3 m.

Tubular heaters are generally satisfactory for local heating where appearance is not important. They are particulary suitable for installing beneath windows so assisting in preventing cold down-draughts.

Units heaters (Fig. 162). Industrial unit heaters consist of a sheet-metal case containing an open-coil element. The case is fitted with adjustable louvres in the front panel and an uncovered opening at the rear. The elements are mounted on insulated cleats and operate at a dull heat. A fan, mounted behind the elements, draws in cool air which passes through the heated elements, and

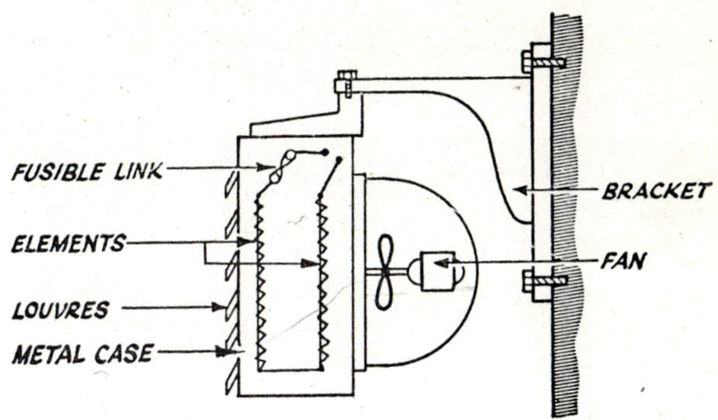

FIG. 162 *Unit heater.*

leaves the heater through the front louvres as a forced convection current of warm air, which quickly penetrates to distances remote from the heater. The element is protected by a fusible link which melts should the fan fail while the heater is in operation. In larger sizes the fan circuit is wired distinct from the heater circuit, and thus the control equipment should be so arranged that it is impossible to switch on the element without first switching on the fan.

The unit is suitable for suspension or wall mounting, and for sizes up to 3kW is designed to operate on single-phase supplies. The larger rating (up to 20 kW) would normally operate on three-phase supplies. The heaters are suitable for thermostatic control, the thermostat being wired in series with the heating element in the smaller sizes and in series with the main contact or coil in the larger sizes.

Fan heaters are also used in domestic installation, and are normally portable. A modern design for this form of heating is the tangential turbine heater which gives an even flow of warm air while the heater is in operation.

Unit heaters are used extensively in domestic, industrial and commercial installations where rapid heating is required.

Oil-filled radiators (Fig. 163). The oil-filled radiator is similar in design to the steam radiator, but is constructed of sheet steel. The radiator contains oil which is heated by a wire-wound element situated in the bottom of the radiator. When the radiator is in operation, the oil adjacent to the element is warmed, rises and is replaced by colder oil remote from the element. The heated oil thus circulates round the radiator, heating the extended fins to an almost uniform temperature.

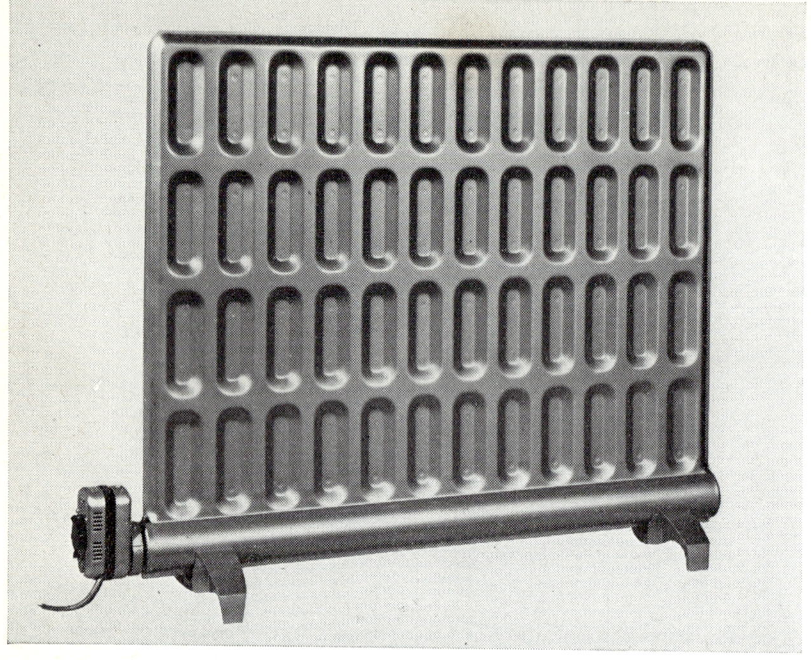

FIG. 163 *Oil-filled radiator.*

The radiator may be of a fixed or portable design with loadings of 1 kW, 2kW or 3 kW, and is very suitable for thermostatic control.

Oil-filled radiators are used mainly for heating small rooms both industrial and domestic, where overall heating is required.

Panel heaters (Fig. 164). Panel heaters may be of the low-temperature type operating at a surface temperature up to 95°C, or of the high-temperature type operating at a surface temperature up to 200°C. One form of low-temperature

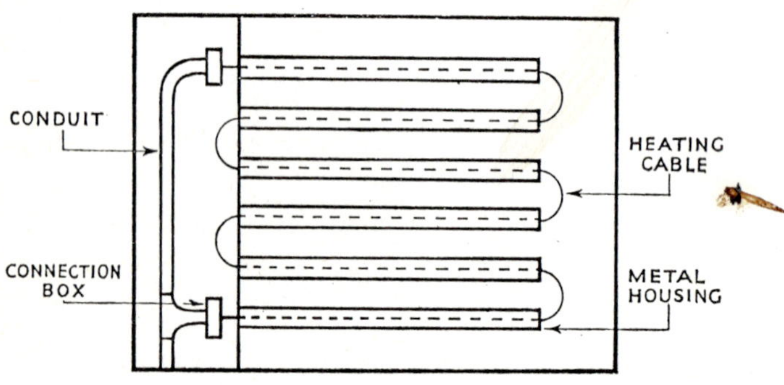

FIG. 164 *Panel heater.*

panel consists of a sheet-steel box, fastened to the wall, the front, which is the heat-radiating unit, being constructed of rolled steel. Immediately behind the front panel is the heating element, consisting of a nickel-chrome wire in the form of a helix, embedded in a refractory material. The element is insulated from the front piece by thin strips of mica, and to prevent heat dissipation from the rear, the back of the refractory units are covered with a layer of insulating material 25 mm to 50 mm thick. The elements operate at dull heat, the loading being approximately 400 W per m^2, giving a surface temperature of approximately 75°C to 95°C.

Panel heaters may be used in any situations in place of convector or oil-filled radiators.

Night-storage heaters (Fig. 165). This is a modern form of panel heater which has become extremely popular, the reason being that they operate on the principle of what is termed thermal storage. The construction of the heater is such that electricity is supplied in the off-peak periods, the heat produced being stored inside the heater and liberated when required, normally during the working periods.

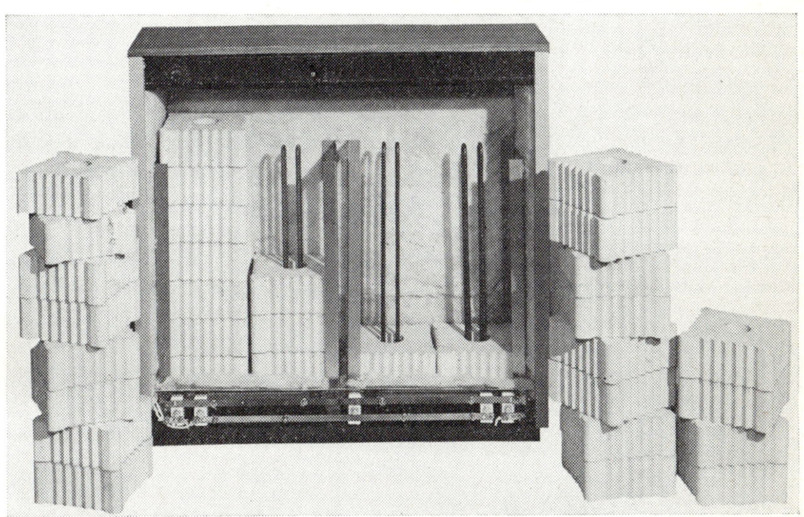

FIG. 165 *Night-storage heater.* (*Courtesy Belling Ltd.*)

There are many types of night-storage heaters, some being fan-assisted, but in the one illustrated in Fig. 165 the elements are wound on refractory panels and surrounded by heat-storage blocks, which are then contained in a sheet-steel case, the inner surface of which is covered with lagging, the case itself containing convection vents to allow the emission of heat. The number of blocks is dependent upon the loading of the heater, normally loadings being from 1 to 3 kW.

Because they operate on the thermal-storage principle, this type of heater is able to take its input energy at the cheap rates offered during off-peak periods. They are therefore used extensively as central heating installations in domestic and commercial installations where the benefit of this type of tariff can be employed.

Underfloor heating. In effect this is a modified form of panel heating, but in this instance the entire floor is used as the heating medium. If the heating units can be embedded in a concrete floor then, as with the night-storage heater, the system may be operated on the thermal-storage principle with the benefit of the off-peak tariff.

Underfloor heating systems are also many and varied, and may be of the rewirable or directly embedded type, operate at low or extra-low voltage. One system consists of a length of floor trunking laid down the centre of the room adjacent to one wall. Metal housing, steel or plastic conduits then radiate outwards, covering the full area to be heated, and being so spaced as to give a loading of 100 to 150 W per m^2 of floor area. The concrete screed is then applied over the conduits or housing, and the heating cable is drawn in and terminated in special connection boxes inside the floor trunking. These connection boxes are supplied from local fuseboards in the normal manner.

The heating cable may consist of copper Nichrome, Kumanol, nickel or other conductors, insulated and sheathed mainly in plastic material. The installation is designed to give a floor temperature of approximately 24°C with an air temperature immediately above the floor of about 17°C. Thermostats are used to control the temperature, and where the system operates on off-peak tariffs, time switches have to be incorporated.

Underfloor heating is used extensively in industrial, commercial and multi-storey blocks of flats where the floor is of a construction suitable for thermal storage. Underfloor heating may also be installed in buildings where the floors are unsuitable for thermal storage, in which case, the installation must be treated as a normal heating installation without the benefits of off-peak tariffs. Also, since such heating is difficult to install in existing premises, it is more usually confined to new buildings.

143. *Describe the construction of a thermostat suitable for controlling a space-heating installation and describe its operation.*

Although there are many types of space-heating thermostats, nearly all operate on the bimetal principle. A bimetallic strip consists of two dissimilar strips joined in parallel together so that when they are heated, the unequal expansion causes the strip to bend in a definite direction, the movement, within certain limits, being proportional to the change in temperature.

In the type used in space-heating installations the movement of the bimetallic strip results in two contacts, wired in series with the heating element, 'making or breaking', according to the rise or fall in temperature. A device is incorporated,

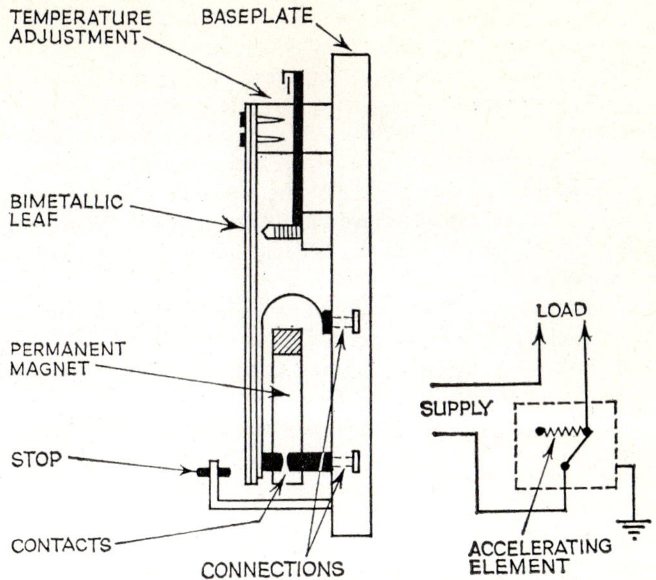

FIG. 166 *Space heating thermostat.*

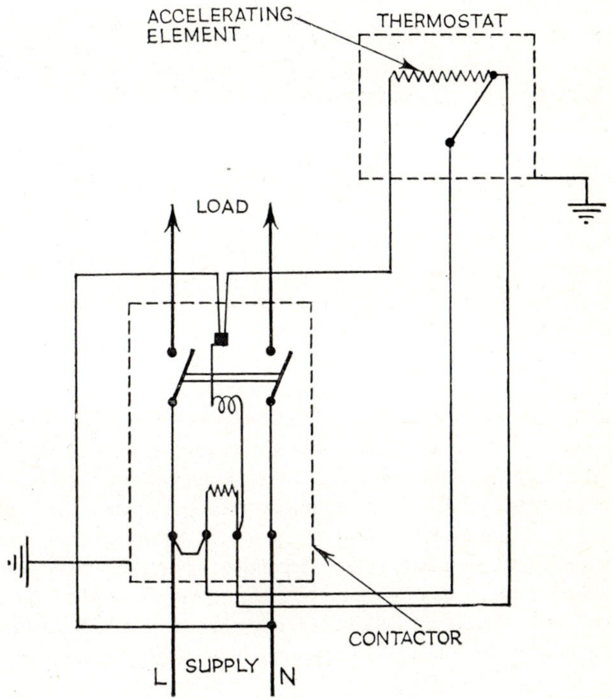

FIG. 167 *Space heating connection for a large heating installation.*

extending beyond the outer case of the thermostat, enabling it to be adjusted so that it will operate within pre-determined temperature limits.

To ensure that the contacts 'make-and-break' quickly and cleanly when the operating point is reached, a magnet may be incorporated to accelerate the contact movement. For low current loadings, a small internal accelerating element may also be fitted which assists in reducing temperature over-shooting caused by thermal delays. This may be omitted when the current is high enough to cause sufficient self-heating to give the accelerating affect. The contacts are electrically connected to terminals in the baseplate which are then connected to the required points in the circuit.

The thermostat position must be carefully sited in order to maintain the correct temperature level throughout the working areas of the room.

Fig. 166 illustrates how such a thermostat is connected with the accelerating element out of circuit, while Fig. 167 shows how the thermostat may be used to control large heating loads, the accelerating element now being incorporated in the circuit.

144. *What factors affect the required loading of a heating installation?*

The quantity of heat required to heat any room depends upon (1) the heat required to raise the room to the required temperature which, in turn, is dependent upon the initial temperature inside the room; and (2) the heat required to maintain this required temperature against the losses incurred through the fabric of the building, that is the walls, windows, ceilings and floors. A rough approximation is to allow 45 W per cubic metre of room space, but a more accurate method is to apply multiplying factors for calculating the losses through the different fabrics.

These factors are found in tables issued by the Institution of Heating and Ventilating Engineers and enable the heat losses of the room to be calculated. In addition to the heat required to overcome the losses, an allowance has to be made for the changing of the air inside the room. Although it is possible to calculate the heat lost through the fabric of the building, it is almost impossible to calculate the exact heat required for the interchanging of the air inside the room. The rate of air change is normally based on the volume of air changed per hour, and normal procedure is to allow one or two air changes per hour. This is purely arbitrary, and may well be exceeded in large garages where the doors are constantly being opened. What should be remembered is that the greater the number of air changes, the higher is the required loading. It must also be remembered that the temperature rise is very largely dependent upon the outside temperature.

A small workshop 10 m by 6 m with a ceiling height of 4 m has a window area of 35 m^2 and a door area of 10 m^2. The workshop is to be heated electrically so that the average temperature is maintained at 20° C when the outside temperature is 2° C.

Calculate the power required in kilowatts on the assumption there will be two air changes per hour.

Density of air $= 1{\cdot}2$ kg/m^3
Specific heat of air $= 1010$ J/kg/deg C.

Heat transmission coefficients in W/m deg C:

Walls, brick $= 1{\cdot}874$
Floor, concrete $= 1{\cdot}136$
Ceiling, plaster $= 2{\cdot}839$
Doors, wood $= 3{\cdot}975$
Window, glass $= 5{\cdot}39$

Area of walls $= (10 \times 4 \times 2) + (6 \times 4 \times 2) - (35 + 10)$

$= 80 + 48 - 45$

$= 83$ m^2

Area of floor $= 10 \times 6$

$= 60$ m^2

Area of ceiling $= 60$ m^2

The heat loss for each item is found from the formula:

heat loss = area × U value × temperature rise

Tabulating

Item	*Area* (*m²*)	*U value* (*W/m deg C*)	*Temp. rise* (*deg C*)	*Heat loss* (*W*)
Walls	83	1·874	18	2800
Floors	60	1·136	18	1227
Ceiling	60	2·839	18	3066
Doors	10	3·975	18	599
Windows	35	5·39	18	3243
		Heat lost through fabric of building		10935

Volume of air $= 10 \times 6 \times 4$

$= 240$ m^3

Mass flow rate $= \dfrac{\text{volume} \times \text{density} \times \text{air changes}}{3600}$

$= \dfrac{240 \times 1{\cdot}2 \times 2}{3600}$

$= 0{\cdot}16$ kg/s

Heat loss	= mass flow rate × specific heat × temp. rise
	$= 0{\cdot}16 \times 1010 \times 18$
	= 2909 W
Total heat required	= 10 835 + 2909
	= 13 744 W
	= 13·74 kW

145. *Describe the construction of a non-pressure water-heater and its operation, and state the situation in which it would be installed.*

The heater (Fig. 168) is of a double-chamber construction, the inner chamber containing the water to be heated, insulated from the outer chamber by means of

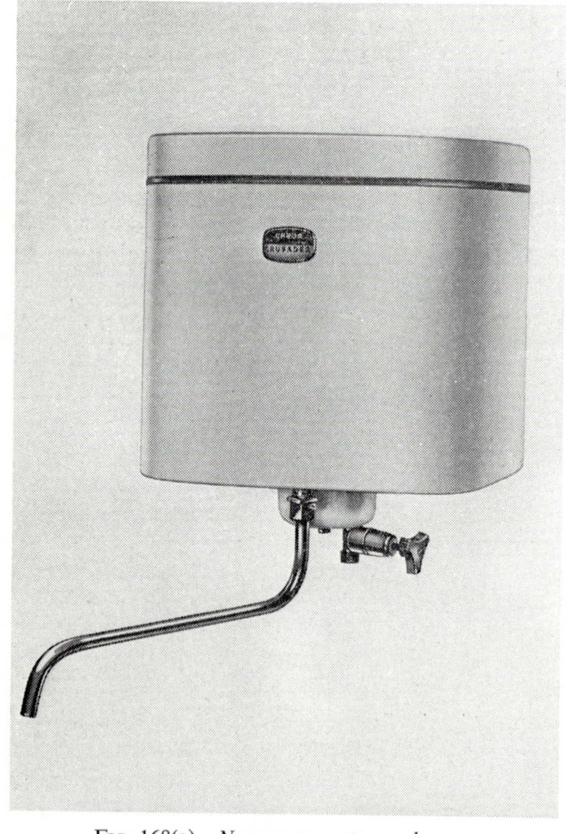

FIG. 168(a) *Non-pressure storage heater.*

lagging, such as granulated cork or glass fibre. Attached to the bottom of the chamber is the immersion heater, the thermostat and the hot-water outlet and cold-water inlet. Provision is made so that the inner chamber is never completely filled with water, a safety valve being fitted to allow for the escape of steam should the thermostat fail to operate satisfactorily. The hot-water outlet is extended upwards towards the top of the inner chamber, a syphoning device being fitted to prevent the dripping of water after the inlet tap has been closed. In many modern storage heaters this device is now unnecessary. A restricting device is also fitted over the inlet, which serves the dual function of preventing the water

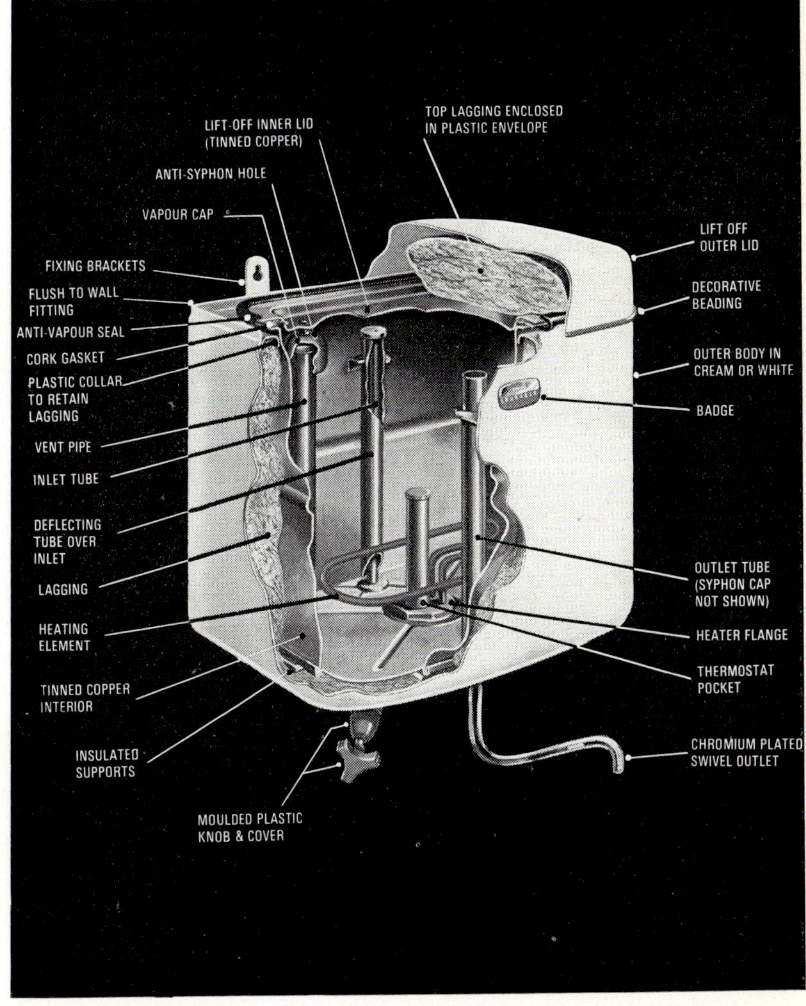

FIG. 168(b)

entering the container from playing directly on the heating element, and also of minimising the pressure so as not to cause mixing of the cold and hot water.

The ends of the heater element are brought out to a terminal block on the base of the inner chamber, and from this block a heat resisting cable is taken to a double-pole switch. A detachable bottom plate then serves to conceal the bare terminals and the necessary interconnecting leads.

The heater is of the wall-mounting type with capacities from 4·5 to 90 litres and ratings from 750 W to 3 kW.

The thermostat controls the temperature of the water. There is no necessity for a tap in the outlet pipe as no water will flow until the tap in the inlet pipe is opened.

This type of heater is, therefore suitable for supplying one point only.

146. *Describe the construction of a thermostat suitable for controlling a water-heater and its operation.*

This type of thermostat comprises an Invar rod fitted inside a phosphor-bronze or brass tube (Fig. 169) all of which is contained within the liquid to be heated. The brass tube is attached to the thermostat head, but the Invar rod is taken inside the head and fastened to the switch mechanism.

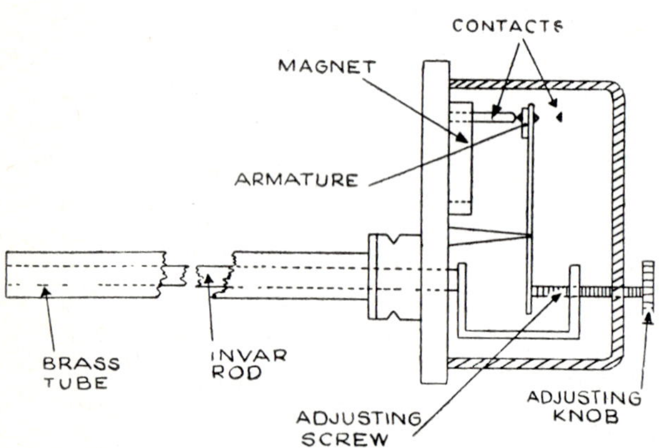

Fig. 169 *Water heating thermostat.*

When the temperature of the liquid rises, the brass tube expands, and the Invar rod, which is attached to the brass tube at the end remote from the head, and which has a negligible temperature coefficient of expansion, is allowed a freedom of movement, resulting in the switch mechanism operating and opening the contacts in the circuit.

When the temperature falls, the brass tube contracts, causing the Invar rod to return to its normal position and the switch contacts to reclose. An adjusting knob outside the head is provided to vary the distance through which the switch has to operate, so allowing the thermostat to operate over a fairly wide temperature range.

A magnet is also incorporated to assist in the 'breaking' operation. The length of the brass tube may vary from 180 mm to 900 mm, although the normal length is 300 mm.

147. *Describe the principle of the electrode-boiler and its advantages and disadvantages when compared to other methods of water heating.*

If an electric current is passed through water, the water acts as a resistance and therefore becomes hot. If the current is uni-directional, oxygen is liberated at the positive electrode and hydrogen at the negative electrode, resulting in a gradual decomposition of the water. It is therefore clear that the electrode-boiler is unsuitable for d.c. supplies.

When connected to a.c. supplies, both oxygen and hydrogen collect at each electrode, but the effect is only slight, and as the quantity of gas liberated is proportional to the current, the higher the voltage at the electrode (consistent with safety), the higher is the efficiency of the boiler.

The electrode water-heater (Fig. 170) therefore consists essentially of a chamber containing water, in which are immersed the required number of electrodes.

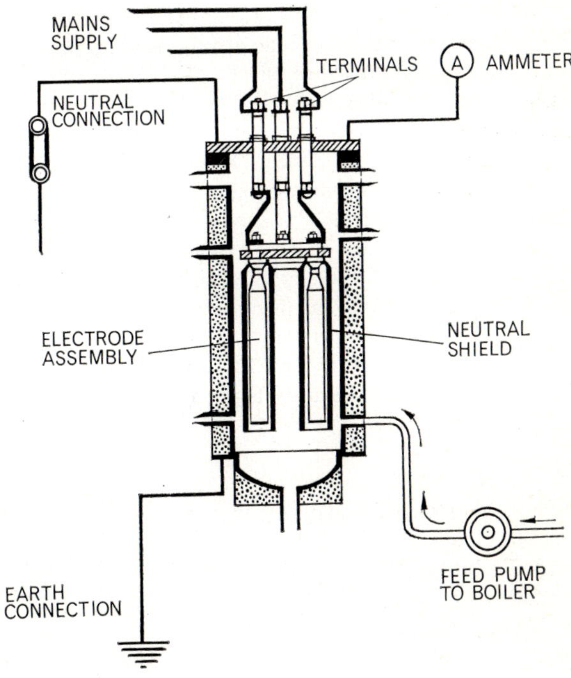

FIG. 170 *Electrode water-heater.*

Dependent upon the requirements, the boiler may be connected to a single-phase or three-phase low-, medium- or high-voltage supply.

The resistance of the water, however, varies with its temperature, so resulting in a variation of current, an undesirable feature, and therefore requiring ancillary equipment to compensate for this variation in load.

One method of compensation is by variation of the conducting path, that is by altering the distance between the electrodes without altering their effective cross-section. Another method is to allow the distance between the electrodes to remain constant and to vary their effective cross-section relative to each other, either by altering the level of water, or by raising the electrodes partially out of the water. A third method, commonly used, is to vary the length of the heating path by varying the position of an insulating shield between the electrodes. Finally, the loading may be varied by incorporating in the external circuit a variable choke or tapped auto-transformer.

One advantage of the electrode water-heater is that for high loadings (in the order of 400 kW or higher) the space occupied by the boiler is far less than that which would be occupied by other water-heating systems. Also, the boiler can operate on high-voltage supplies with a corresponding increase in efficiency. Another advantage is that the electrode-boiler cannot burn out. Should the heater boil dry, the circuit is automatically broken, so reducing the risk of fire.

The disadvantage of the electrode-boiler is that where the loading is only of a few kW rating, the automatic compensating gear and the electrode construction take up far more space than that of an equivalently rated immersion heater and thermostat, and is not so economical to install.

Also, because of safety considerations, the Regulations concerning the installation of electrode water-heaters are very stringent, and permission must be obtained from the Postmaster General before low- and medium-voltage electrode boilers may be put into commission.

Where the thermal storage principle is adopted by lagging the storage tanks to conserve the heat, the advantage of the off-peak tariff can again be obtained. The boiler is thus connected to the supply during the off-peak periods, heat being conserved in the water in the heating chamber, which is then allowed to circulate round the radiators during the normal working period.

Although the running costs are higher with an electrode-boiler installation than one using solid fuel, it has the advantage of being clean, easily sited, and simple to control.

148. *Describe the construction and operation of a heat-controlled iron.*

The automatically heat-controlled electric iron (Fig. 171) contains a heating element which consists of three mica sheets, the centre one being wound with a nickel-chrome heating element. The mica sheets are placed above a relatively thick iron plate, the sole of which is highly polished. On top of the mica sheets is placed a sheet of asbestos which separates the mica from a solid iron block which

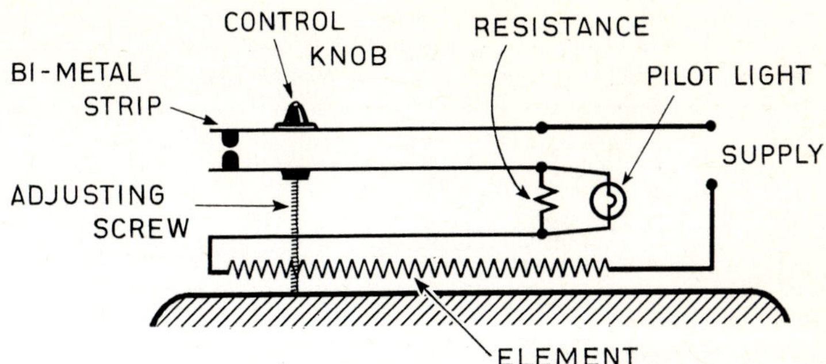

FIG. 171 *Heat-controlled iron.*

holds the element in position. A bimetallic type of thermostat is also fitted inside the iron, the contacts of which are connected in series with the element and also a small resistor, across which a small pilot lamp is connected. The value of the resistor is such that a potential of 2·5 V is obtained across the resistor and the pilot lamp. The lamp is usually fitted in the handle, and it is illuminated when the thermostat contacts are closed and the element is connected to the supply.

The distance through which the bimetallic strip must move before opening or closing the contacts is adjustable, a control knob being fitted on top of the outer cover for this purpose.

With this type of iron, the flexible cord is connected directly to terminal pillars inside the iron so producing a more positive earthing action.

An adaptation of the heat-controlled iron is the steam iron which contains a chamber into which water is poured. When the element is sufficiently heated, the water is turned to steam, which passes through small holes in the sole plate when the iron is set to steam control.

149. *Describe how simmerstat control is applied to the boiling ring of an electric cooker.*

If a boiling plate is switched on, say for ten seconds and then swiched off for ten seconds, the heating effect is virtually the same as if the element was switched on at half value for twenty seconds. If, therefore, the ratio of the on-off times can be suitably adjusted, a wide range of varying heat outputs may be obtained. Simmerstat control enables this to be achieved.

The control unit (Fig. 172) for such a system comprises a bimetallic strip which is heated by a small heating element. The movement of the bimetallic strip, the amount of which may be varied by means of the control knob, opens a pair of contacts connected in the heating element circuit. When the bimetallic strip cools, it resumes its original position, the contacts close again. The cycle of operation is repeated for as long as the circuit is connected to the supply.

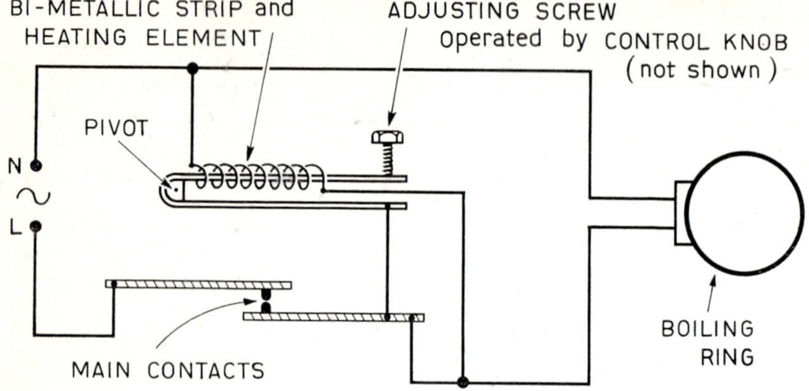

FIG. 172 *Simmerstat control of a boiling ring.*

This means that the supply to the boiling plate element follows the same pattern as that of the bimetallic strip, so that its heat output can be adjusted over a very wide range.

The control knob has several settings, the OFF position being just below the bottom setting. One advantage of this type of control is that the element may consist of one resistor only and, therefore, be of a more robust design than the normal element.

150. *What are the main causes of breakdown in electrical heating appliances and how may they be located and repaired?*

Electric fires. After a period of time, a fire element tends to become brittle and break, and as it has usually become less efficient because of a tendency for the fireclay to blacken, it is preferable to replace the complete element rather than to try and re-wind it.

Another source of trouble with the parabolic type of fire is arcing at the terminal connections. This should be attended to immediately, as constant arcing produces a high resistance contact. Cleaning of the element wire and the terminal connections are usually all that is necessary.

Another source of trouble is where the flexible cord is connected to the fire. To minimise the risk of damage by heat, the flexible cord should be protected by heat-resisting sleeves or beads. It is not generally realised that the earth-continuity conductor in the flexible cord may become broken and remain unnoticed until someone receives a shock from the appliance. Only periodic testing will prove that earthing is efficient.

Convector, tubular and panel heaters. Because the elements operate at a much lower temperature than those of the radiant heater, they are not so prone to breakdown. All types of switch, however, occasionally cause trouble and should be repaired or replaced when necessary. Thermostats are better replaced, and the faulty thermostat should be returned to the manufacturer for overhauling and recalibration, unless the fault is due to 'welding' of the contacts which can be cured *in situ.*

Fan heaters. The most probable cause of trouble is the failure of the fan to maintain its correct speed, resulting in a possible overheating of the heating elements. To avoid this a fusible link is normally incorporated in the heater element circuit. This is designed to melt when the temperature exceeds a predetermined value. It is also necessary to ensure that the air inlet is not obstructed, so restricting the cold air flow into the heater. The electric hair-dryer also operates on the same principle as the fan heater and is therefore subject to the same troubles.

Underfloor heating. In both the directly-embedded and the rewirable type, part of the system may break down possibly because of the screed curling, through faulty laying of the cable, or even from mechanical damage, when the floor is being opened at some later period.

Some form of signal generator has to be used for plastic-insulated heating cables, and 'hot-spot' generators with visual or audio detectors for m.i.c.c. cables. By these methods, the fault is quickly detected, and the floor opened up to make the necessary repair. It is usually unnecessary to open large areas of the floor, as the fault-tracing method enables the approximate position of the fault to be pinpointed. With the rewirable system, it should be possible to draw in a fresh cable, although it has been known for the heating cable to become welded to the tube, in which case the floor has to be opened to rectify the fault.

Water heaters. Should the heating elements of the withdrawable type of heater fail, it is a simple matter to take out the faulty element and replace it, without taking out the complete heater. Where the heater is not of the withdrawable type, the heater must be taken out completely and replaced, should the element fail. A fault that sometimes occurs is where the wrong type of cable has been used inside the heater head. The insulation chars and eventually short-circuits, blowing the fuse. The flexible cord should be replaced with the correct type of flexible cord. For a.c. supplies the controlling switch is of the micro-gap type, and must be double-pole so that both poles of the supply are disconnected whilst maintenance work is being carried out.

Electric cookers. The elements of electric cookers are normally of the plug-in type, so that if one fails, the simplest method is to take out the complete boiling ring, etc., and return it to the maker for repair. Boiling rings are provided with adjusting screws so that the plate may be levelled after it has been put into service.

Faulty elements may be replaced in existing fire-bars in grill boilers. Oven elements may also be replaced where these are of the open-element type. As replacing a fuse normally entails removing some of the framework, the main control should be opened before a fuse is replaced. If three-heat switching is used, the contacts may be tested with a test lamp, but the circuit wiring must be disconnected, otherwise misleading results might occur through feed-backs. Unless the fault is caused by welding of the contacts, faulty thermostats and simmerstats should be replaced by new ones, and the faulty one returned to the makers for repair and calibration.

Electric irons. Repairs to the non-automatic type are very simple and usually only entail fitting a new element or repairing or renewing a faulty flexible cord. The element must be held firmly in position by the iron plate and the asbestos pad fitted between the element and the plate.

Repairs to the heat-controlled iron are more difficult to carry out since the entire assembly has to be dismantled to replace a faulty element. This entails readjustment of the thermostat by means of a nut and screw projecting through the thermostat assembly, and several adjustments may have to be made before the thermostat is correctly calibrated. Although the operation of the thermostat can be checked by watching the pilot light, this is not a reliable guide, and, therefore, the calibration should be checked against the temperature of a testing device, manufactured specially for this purpose.

Electric kettles. Occasionally the ejector device fails to disconnect the kettle from the supply when the kettle has boiled dry, so burning out the element. This may be caused by corrosion or stiffness of the connector plug in its socket, and. therefore, this should be periodically checked. Should the element be faulty, it may easily be replaced by unscrewing the metal shroud surrounding the contacts, and lifting the entire element out, either through the kettle lid or from underneath. Washers are provided to prevent water leaking through the connector hole, and should they become damaged, they should be replaced by new ones. Kettle connectors are normal circular, so ensuring that the spring clips connected to the earth-continuity conductor make a sound connection to the metal shroud. When fitting a new connector, only one of an approved kind should be fitted as a sound earth connection is a vital necessity.

Summary. No matter what appliance is connected to the supply, it is essential to ensure that the flexible cord is of the correct size and type for the condition under which it is to be used, and also that the protective device is rated correctly for the cables it is intended to protect. Also, as the appliance will continue to function even though the earth-continuity conductor is faulty and perhaps remain unnoticed, it is essential that the appliance is frequently tested, possibly on the appliance tester of a loop-impedance tester.